Nanoscopic Materials
Size-dependent Phenomena

Nanoscopic Materials
Size-dependent Phenomena

Emil Roduner
Institute of Physical Chemistry, University of Stuttgart, Stuttgart, Germany

RSCPublishing

The front cover image was designed by Ludovico Cademartiri and illustrates a variation in properties of nanocrystals dependent on their size.

ISBN-10: 0-85404-857-X
ISBN-13: 978-0-85404-857-1

A catalogue record for this book is available from the British Library

Published by The Royal Society of Chemistry,
Thomas Graham House, Science Park, Milton Road,
Cambridge CB4 0WF, UK

Registered Charity Number 207890

For further information see our web site at www.rsc.org

Typeset by Macmillan India Ltd, Bangalore, India
Printed and bound by Henry Ling Ltd, Dorchester, Dorset, UK

Preface

When we look back over historical times we have a feeling that the changes in human society, life style and culture accelerate more and more the closer we approach the present time. This is a very intuitive perception since we have no objective measure of the rate of change, and it may be biased to some extent by the fact that more information in general and therefore also more information about the rate of change is available for the more recent centuries. Should changes indeed accelerate exponentially or even faster then they extrapolate to infinity. Nothing though goes to infinity in nature, so there must be a term that slows down this rapid development and turns it into a critically damped or oscillatory behaviour. This decelerating term may at present not be visible, but its likely presence makes extrapolations into the future extremely difficult.

One of the things which are expected to rush over our globe and change our lives in near future is *nanotechnology*, often termed a key technology of the 21st century. The rate and the extent of the prospected changes are horrendous. The prospects for future generations are therefore loaded with hypes, hopes and fears, and they are difficult to assess reliably. Probably the best thing that we can do to cope with this uncertain future and reduce fears and risky developments is educating as many people as possible to enable them to establish their own and independent judgement. This is the main aim of the present book.

In its strict sense the term nanotechnology is used for methods which permit the manipulation and controlling of *individual* atoms, molecules, or other entities of matter on a nanometer scale. The technology of making and controlling things on a small scale has been in the focus of much of the work during the past decade. Several books have come out on technological aspects, while understanding things has been somewhat in the second row, at least in the chemical discipline. The latter is the more specific focus of the present text.

In a broader sense the term *nanotechnology* includes the application of *nano-materials*. Nanomaterials are not the products of nanotechnology in its strict sense; rather they represent large amounts of materials in a size range of 1-100 nm which because of their small size adopt a wide variety of properties different from those of the bulk. They are in the focus of materials science which has undergone a veritable revolution over the past decades. Fine powders, compacted or sintered, block-copolymers, organic-inorganic nanocomposite materials, carbon nanotubes and inorganic porous materials derived from sol-gel processes, self-assembled dots and layers, thin films and fibres have found a multitude of applications. Most of these materials have a surface or interfacial area which can be as high as 1000 m^2 per gram and more. Surface effects are no longer negligible; rather, they often dominate many of the properties of practical relevance. They have to be understood in order to permit the manufacturing of materials with tailored properties and sufficient stability. Modern manufacturing technologies aim at mass production, and the interest is in the *collective* rather than the individual property.

Physical chemistry is the science at the intersection between physics, chemistry, and engineering. And yet physical chemistry textbooks around the world have nearly not taken notice of the special size-dependent phenomena of nanosize materials. We continue to teach chemistry students in thermodynamics all possible aspects of *volume work* of ideal gases – things of relevance in combustion and flight engineering. For materials scientists and people involved in nanotechnology it would be much more important to get at least a basic understanding of *surface work* and of other phenomena near interfaces. The present text is intended as a contribution to fill this gap. It wants to introduce this matter on a phenomenological basis, it reports quantitative relations (scaling laws) where available, and it leads the way to original literature where further details may be found.

Many of the basic concepts are not new. Capillary effects and the Gibbs-Thomson equation, for example, have been known for over a century. However, technological progress in recent decades has led to far better control of nanoscale materials. This in turn has presented a challenge on theory and has led to refinements of the original concepts. Moreover, the availability of well defined nanomaterials with new and tailored properties has led to stimulating promises and exciting hopes. Revolutionary electronic elements and devices such as quantum computers are some of the present challenges which are in the focus of many scientists. Already, specially designed nanoparticles are being used as carriers for drugs which permit their distribution in the human body and a controlled release at the target site. On the other side of the medal we have new potential risks which at this point have been largely unexplored. All these prospects have imposed a new weight on the concept of size-dependent phenomena in fundamental and applied research and even in every day life. In order to warrant further progress and success it is essential that these concepts become part of the active, every-day thinking of young people, and it is hoped that this book will contribute to it.

Nanotechnology is a rapidly developing field. A large number of scientists is working hard to push the frontiers further and turn dreams into reality. New publications with exciting new results appear day-for-day. Any book about nanotechnology has to catch the subject on the fly. Since the focus of the present work is on the fundamental principles rather than on what we can do today it is nevertheless hoped that a large fraction of the material covered here will remain valid for some time.

Scientific progress at the frontier of a new field is based entirely on the creativity of the involved scientists. Seminal scientific work therefore needs the freedom to explore what scientists come up with in their dreams. Programmed research may be suitable to achieve a goal in a collaborative effort, but no matter who has defined the program it will never have the flexibility to adapt to such dreams and therefore it will never be creative to the same extent as free research. However, free research does not mean that it is free of responsibility for what it inherits to society and to nature. There is a *personal responsibility of every scientist* to comply with accepted ethical guidelines. It is extremely damaging to the image of science when this is not respected. When scientists or companies value personal benefits higher than benefit to society and nature, then politics will have to set limits. It is quite certain though that specific and appropriate laws will always lag far behind any scientific developments. Law-makers will try to cope with this situation by increasing the rigidity of the limits, which of course will have the effect that also harmless developments are cut. It is the aim of the last chapter to extrapolate some of the developments into the future and to address some of the ethical bounds involved.

This text has developed out of lecture notes for a graduate level course at the University of Stuttgart. Thanks are due to the students who have served as "guinea pigs" and endured the first attempts patiently or contributed with valuable comments. Gratefully acknowledged are also the contributions of Bettina Beck, Stefan Karolczak and Kurt Hofer who have cross-read the text and commented on its readability and content, of Diana Zauser and Gabriele Bräuning who have assisted with figures, Isolde Rosenkranz who has coped with copy right subjects, and Inge Blankenship who has produced the subject index. Last but not least I would like to thank my dear wife Hanny and my family for their continuous encouragement and support of this work.

Emil Roduner
Stuttgart, November 2005

Contents

Chapter 1 Introduction   1
   1 Clusters and Nanoparticles   1
   2 Feynman's Vision   2

Chapter 2 Bulk and Interface   5
   1 Gradients Near Surfaces   5
   2 The Coordination Number Rules the Game   6
   3 Surface Science, a Source of Information for
      Nanoscience   8
   4 Particle Size and Microstrain   11
   5 Biomimetics: Nature as a Source of Inspiration for
      Strategies in Nanotechnology   17

Chapter 3 Geometric Structure, Magic Numbers, and Coordination
Numbers of Small Clusters   21
   1 The Consequences of the Range of the Radial Potential
      Energy Function   21
   2 Magic Numbers by Geometric Shells Closing   26
   3 Magic Numbers by Electronic Shells Closing 29
   4 Cohesive Energy and Coordination Number   34

Chapter 4 Electronic Structure   41
   1 Discrete States Versus Band Structure   41
   2 The Effects of Dimensionality and Symmetry in
      Quantum Structures   42
   3 The Nonmetal-to-Metal Transition   47
      3.1 General Criteria   47
      3.2 The Special Case of Divalent Elements 49
      3.3 Experimental Criteria of Metallic Behaviour 51

4   Work Function, Ionisation Potential and Electron
     Affinity   55
5   Electronic Structure of Semiconductor and Metal
     Clusters   60
     5.1   Optical Transitions in Semiconductor Nanoclusters   60
     5.2   Photochemical and Photophysical Processes of
            Semiconductor Nanoparticles   65
     5.3   Optical Properties of Metal Nanoclusters   69
6   A Semiconductor Quantum Dot Electronic Device   74

Chapter 5   Magnetic Properties   81
1   A Brief Primer on Magnetism   81
     1.1   The Basic Parameters   81
     1.2   Curie Paramagnetism   82
     1.3   Curie–Weiss Paramagnetism   83
     1.4   Antiferromagnetism   84
     1.5   Ferromagnetism and Ferrimagnetism   84
     1.6   Molecular Magnets   86
     1.7   Superparamagnetism   88
     1.8   Other Forms of Magnetism   90
2   The Concept of Frustration   91
3   Magnetic Properties of Small Clusters   95
     3.1   Theoretical Predictions   95
     3.2   Experimental Observations of Magnetism in
            Clusters   100
4   Ferromagnetic Order in Thin Films and Monoatomic
     Chains   106
5   Finite Size Effects in Magnetic Resonance Detection   109
     5.1   Nuclear Magnetic Resonance   109
     5.2   Electron Spin Resonance   111

Chapter 6   Thermodynamics for Finite Size Systems   119
1   Limitations of Macroscopic Thermodynamics   119
     1.1   A Formal Approach   119
     1.2   Systems Beyond the Thermodynamic Limit   120
     1.3   The Breakdown of the Concept of Phases   122
2   The Basics of Capillarity   124
3   Phase Transitions of Free Liquid Droplets   128
4   The Lotus Effect   129
5   Classical Nucleation Theory   136
6   Shape Control of Nanocrystals   141
7   Size Effects on Ion Conduction in Solids   148
8   Principles of Self-Assembly   152

Chapter 7 Adsorption, Phase Behaviour and Dynamics of Surface Layers and in Pores 163
  1 Surface Adsorption and Pore Condensation 163
    1.1 The Langmuir Adsorption Isotherm 163
    1.2 The Brunauer–Emmett–Teller (BET) Equation 163
    1.3 Adsorption in Micropores 166
    1.4 Adsorption and Condensation in Mesopores 168
    1.5 Determination of Mesopore Volumes and Mean Pore Size 169
  2 Adsorption Hysteresis and Pore Criticality 170
  3 The Melting Point of Pore-confined Matter 178
  4 Layering Transitions 185
    4.1 Layering of Solids and Liquids Adsorbed on Smooth Surfaces 185
    4.2 Layering Transitions of Confined Fluids in Smooth Pores 187
  5 Liquid Coexistence and Ionic Solutions in Pores 191
  6 The Effect of Pressure 193
  7 Dynamics in Pores 194
    7.1 Dielectric Properties 194
    7.2 Diffusion and Viscosity Under Confinement 198

Chapter 8 Nucleation, Phase Transitions and Dynamics of Clusters 209
  1 Melting Point and Melting Enthalpy 209
    1.1 Introduction 209
    1.2 Supported Tin Clusters 210
    1.3 Melting of Cadmium Sulfide Nanocrystals 214
    1.4 Free Sodium Clusters 214
    1.5 Isolated Silver Clusters 219
    1.6 Simulated Melting Behaviour of Further Metal Clusters 221
    1.7 Discrete Periodic Melting of Indium Clusters 221
    1.8 Hydrogen-Induced Melting of Palladium Clusters 222
  2 Dynamics of Metal Clusters 223

Chapter 9 Phase Transitions of Two-Dimensional Systems 233
  1 Melting of Thin Layers 233
  2 Structural Phase Transitions in Thin Layers 233
  3 Glass Transition of a Polymer Thin Film 235
  4 Surface Alloy Phases 236

Chapter 10 Catalysis by Metallic Nanoparticles   239
   1   Some General Principles of Catalysis by Nanoparticles   239
   2   Size-Controlled Catalytic Clusters   241
   3   Shape-Dependent Catalytic Activity   246
   4   The Effect of Strain   248
   5   The Effect of Alloying   252
   6   Metal-Support Interaction   255
   7   The Influence of External Bias Voltage   257

Chapter 11 Applications: Facts and Fictions   263
   1   Nanomaterials   263
       1.1   General Considerations   263
       1.2   Applications in Medicine   263
       1.3   Intelligent Surfaces   265
       1.4   Applications in Catalysis   265
       1.5   Applications in Environmental Technologies   265
   2   Nanotechnology   266
       2.1   Applications to Nanomechanics   267
       2.2   Applications in Nanoelectronics   269
       2.3   Applications of Single Spin- and Nanomagnetism 272
       2.4   Applications of Optical Properties   273
   3   Hopes, Hazards and Hype   275
       3.1   Is Nanotechnology Useful?   275
       3.2   Potential Health and Environmental Hazards   276
       3.3   Ethical and Social Threats from Nanotechnology 276
       3.4   Is Nanotechnology but Hype?   278

Subject Index   281

CHAPTER 1
Introduction

1 Clusters and Nanoparticles

Particles with a size between 1 and 100 nm are normally regarded as *nanomaterials*. Figure 1 shows the size of nanoparticles in comparison with other small particles. In general, nanomaterials may have globular, plate-like, rod-like or more complex geometries. Near-spherical particles which are smaller than 10 nm are typically called *clusters*. The number of atoms in a cluster increases greatly with its diameter, demonstrated in Figure 2 for sodium clusters. At 1 nm diameter there are 13 atoms in a cluster and at 100 nm diameter the cluster can accommodate more than 10^7 atoms. Clusters may have a symmetrical structure which is, however, often different in symmetry from that of the bulk. They may also have an irregular or amorphous shape. As the number of atoms in a cluster increases, there is a critical size above which a particular bond geometry that is characteristic of the extended (bulk) solid is energetically preferred so that the structure switches to that of the bulk.

It is below a dimension of 100 nm where properties such as melting point, colour (*i.e.* band gap and wavelength of optical transitions), ionisation potential, hardness, catalytic activity and selectivity, or magnetic properties such as coercivity, permeability and saturation magnetisation, which we are used to

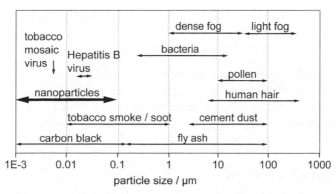

Figure 1 *Typical size of small particles. The regime below 0.1 μm corresponds to the dimension of nanoparticles*

1

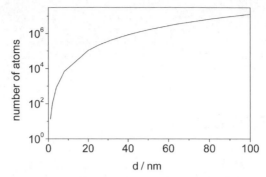

Figure 2   *Number of sodium atoms in a spherical cluster of diameter d*

thinking of as constant, vary with size. We basically distinguish two types of variations as a function of size:

- *Scalable effects*: Surface atoms are different from bulk atoms. As the particle size increases, the surface-to-volume ratio decreases proportionally to the inverse particle size. Thus, all properties which depend on the surface-to-volume ratio change continuously and extrapolate slowly to bulk values.
- *Quantum effects*: When the molecular electronic wave function is delocalised over the entire particle then a small, molecule-like cluster has discrete energy levels so that it may be regarded like an atom (sometimes called a *super atom*). The simplest model for it is that of a particle in a box. Adding more atoms to the cluster changes the size of the box continuously so that the energy levels close up to some extent. More importantly, adding more atoms means adding more valence electrons to the system. Thus, whenever a shell of sometimes multiple degenerate energy levels is filled the next electron has to be accommodated in the next shell of higher energy. The situation is analogous to the evolution of properties with increasing atomic number in the periodic table. Filled shells represent a particularly stable configuration. Properties such as ionisation potential and electron affinity are well known to display a discontinuous behaviour as one moves along the periodic table. For clusters consisting of atoms with strongly overlapping atomic orbitals, *i.e.* for metals and semiconductors, the situation is analogous.

Quantum effects are more pronounced for small clusters and often superimposed on a smoothly varying background of a scalable effect. Clusters are interesting intermediates between single atoms and bulk matter and represent a natural laboratory to 'see both ends from the middle'.

2   Feynman's Vision

On 29 December 1959, at the annual meeting of the American Physical Society, Richard Feynman addressed the audience with his visionary and by now historical and legendary lecture under the title – *There is Plenty of Room at the Bottom: Invitation to Enter a New Field of Physics.*[1] With this talk on the problem of *manipulating things on a small scale*, Feynman opened the field of nanotechnology. Today, more than four decades later, the field is finally seen to really take off. It is amazing how closely some of the key developments follow Feynman's vision. *Why cannot we write the entire 24 volumes of the Encyclopaedia Britannica on the head of a pin?* he asked. We know how successful information technology has been in its work towards this goal and we should be aware of how much this has influenced our lives. *Make the electron microscope hundred times better*, Feynman said. The development of the atomic force microscope was one of the milestones on the way not only to observe but also to manipulate in atomic dimensions. Amazingly, Rohrer and Binnig achieved this goal with their cantilever-based instrument in a single step, rather than by a hierarchy of smaller and smaller robots – *training an ant to train a mite* – as Feynman suggested. In 1986, the two scientists were honoured for their achievement with the Nobel Prize in Physics. It is quite obvious today that the invention of the scanning tunnelling microscope finally triggered the boom in nanotechnology to which the direct observation of very small scale structures down to individual atoms is essential. Obviously, seeing things directly is more convincing for vision-based beings than just having measurements which are in agreement with a model.

Some of the inspiration came from biology. Feynman understood that information is stored on a molecular level in biology, that cells manufacture substances and operate on a small scale, that the human brain is a wonderful and efficient miniaturised computer. *Consider the possibility that we too can make a thing very small which does what we want, an object that manoeuvres at that level*, he suggested. While his talk was primarily technology oriented, he knew that physics, chemistry, biology and engineering are all relevant and must all be involved. He realised that making things smaller was not just a technological problem of scaling down. He saw that certain things changed principally. Magnetism, for example, is a cooperative phenomenon and involves domains which cannot be reduced down to an atomic size. Atoms differ from the bulk in their quantum nature. Most significantly, Feynman predicted that *when we have some control of the arrangement of things on a small scale we will get an enormously greater range of possible properties that substances can have*. It is exactly the arrangement of things on a small scale which is the foundation for all the excitement about nanomaterials and for the success of modern materials science.

References

1. R.P. Feynman, *Eng. Sci.*, 1960, 23, 22; *J. Micromech. Syst.*, 1992, 1, 60; www.zyvex.com/nanotech/feynman.html.

CHAPTER 2
Bulk and Interface

1 Gradients Near Surfaces

A simplistic view distinguishes between gaseous, liquid, and solid phases of a single chemical component, and it assumes that there is a sharp phase boundary where properties change discontinuously between two homogeneous neighbouring phases. In nature there are of course no discontinuities, and when we focus our view onto the interface we first find that it is corrugated on a level of atomic or molecular dimensions. More importantly, although the chemical potential of an equilibrated system is by definition the same in the bulk of both phases and does not change through the phase boundary, most other properties depend on the phase, but they change continuously along the coordinate perpendicular to the interface, often with superimposed oscillations. An example is given in Figure 1.

Let us assume that we cleave a perfect single crystal in vacuum. After cleavage, the density drops to zero instantaneously at the new interface, and the atoms or molecules near the surface experience an asymmetric interaction because some of the partners on one side have disappeared. Thus, they have to adjust to find a new balance of interacting forces. This is achieved by relaxation of the lattice. As a consequence, the local lattice parameters change, and the

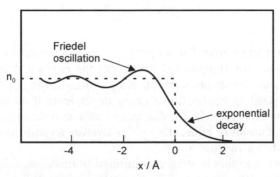

Figure 1   *Schematic drawing of the charge density distribution (Friedel oscillations) across the surface of a spin-polarised electron gas representing the conduction electrons of a metal*
(Adapted from Ref. 1, see also Refs. 2–4, used by permission of Wiley)

density becomes a function of distance from the surface. It is plausible that this gradual change in density is paralleled by changes in most other properties. The wave function is modified, bond lengths, bond strengths, molecular orientation, atomic or ionic mobility, conductivity, and even the index of refraction change. Essentially, as we approach a phase boundary, there is a finite gradient of all properties which may be of interest to us. Note, however, that quite often interfaces are nonequilibrium metastable regions so that not even the chemical potential is constant across a phase boundary.

The length scale λ_0 over which these gradients extend depends on the range of the interactions between the particles. For example, the cohesion of noble gas atoms is governed by van der Waals forces. They have an attractive contribution that decays approximately with r^{-6}, thus it has a relatively short range of a few atomic diameters. The Coulomb interaction in an ion pair goes as $1/r$, it is thus in principle of much longer range. However, in an ionic lattice the alternation of anions and cations leads to an effective Coulomb potential which decays as $r^{-5.5}$

As long as a system has dimensions which are large compared to λ_0 its behaviour is determined by its bulk properties, but when its diameter reaches values as low as a few times λ_0, then its properties become dominated by those of the boundary region. Such a system is not strictly periodic, but it is also not homogeneously amorphous, rather the surface is always different from the interior so that there is a radial gradient in all properties.

2   The Coordination Number Rules the Game

An important parameter for the description of size-dependent phenomena is the *surface-to-volume ratio, A/V*. For spherical particles of diameter d (radius R) we have

$$\frac{A}{V} = \frac{4\pi R^2}{4\pi R^3/3} = \frac{3}{R} = \frac{6}{d} \tag{1}$$

The d^{-1}-dependence holds for simple geometries such cubes, long cylinders or thin plates, but for complicated structures the relation is less straightforward. Equilibrated matter often adopts simple shapes. Many properties therefore obey to a good approximation a linear dependence if plotted against d^{-1}. For spherical or cubic particles the diameter scales with the inverse third power of the number of atoms or molecules, $N^{1/3}$, therefore an equivalent straight-line plot is obtained as a function of $N^{-1/3}$.

Another expression that is often encountered in this context is the *dispersion F*, which designates the fraction of atoms in the surface shell of a material. For cubic clusters with n atoms of radius r_0 along the edge the total number of atoms is $N = n^3$, the number of atoms at the surface is given by $6\,n^2$ for the six faces, corrected for double counts of the 12 edges ($12\,n$) and reinstalling the

8 corners, so that the dispersion becomes

$$F = \frac{6n^2 - 12n + 8}{n^3} = \frac{6}{N^{1/3}}\left(1 - \frac{2}{N^{1/3}} + \frac{8}{6N^{2/3}}\right) \approx \frac{6}{N^{1/3}} \tag{2}$$

This behaviour is illustrated in Figure 2.

For large spherical clusters of radius R the dispersion is proportional to the volume of a shell of thickness $2r_0$ divided by the total volume. Assuming the same packing density and recognising that $N = R^3/r_0^3$ the dispersion becomes

$$F \approx \frac{4\pi R^2 \times 2r_0}{4\pi R^3/3} = \frac{6r_0}{R} = \frac{6}{N^{1/3}} \tag{3}$$

which is the same as from Equation (2). Accounting for the fact that the packing density at the surface is lower, other sources use $F \approx 4/N^{1/3}$.[6] The above expression is applicable for N larger than about 100, for smaller clusters detailed structural information is required. On this basis we obtain $F = 0.4$ for $N = 10^3$, and $F = 0.04$ for $N = 10^6$.

The key parameter which permits a more precise interpretation on a molecular level of the effects described by the surface-to-volume ratio and the dispersion is the number of direct neighbours, the *coordination number*. Atoms or molecules at and near the surface, and even more at edges and corners, have fewer neighbours and are therefore less strongly bound than those in the bulk. This is the reason why the surface has a higher energy, why it often melts first, and why it affects many other properties of the particle. It will be come clear throughout this text that a large number of size-dependent phenomena can be explained on the basis of the coordination number. This is the case whenever a

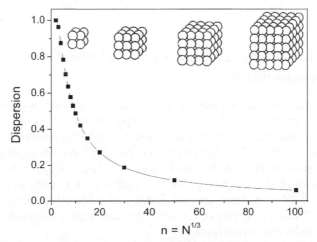

Figure 2 *Evolution of the dispersion F as a function of n for cubic clusters up to n = 100 (N = 10^6). The structure of the first four clusters is displayed*

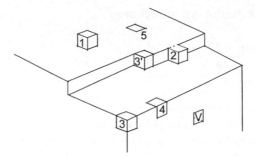

Figure 3 *Crystal built up of cubic units to introduce the various sites. 1: surface adatom,*
2: step adatom, 3: corner atom, 3': kink atom, 4: edge atom, 5: surface atom, V:
surface vacancy. The label number gives also the first shell coordination number

property scales with the inverse particle diameter, or equivalently, with the
inverse cube root of the number of atoms in a system.

The term coordination number is often explained based on a simplified
model crystal built up of cubic atoms (Figure 3). It is the number of nearest
neighbours which are bound via a face of the cube. Each such neighbour
contributes with a certain binding energy to the stabilisation of an atom. The
most stable atom is one in the bulk (not shown) which has six such neighbours
(12 for close packed hard spheres); the least stable is a surface adatom which
has only one. The numbers in Figure 3 are at the same time the coordination
numbers. They are a first measure of the energy of the atom at a certain site,
which is of importance in crystal growth and in catalysis.

In a better approximation the neighbours which are bound via an edge and
those in contact via a corner would also contribute to stabilisation, but to a
lesser extent. This would also distinguish between the energy of a corner atom
(3) which has three face neighbours and three edge neighbours, and that of a
kink atom (3') which has also three face neighbours but six edge neighbours.

3 Surface Science, a Source of Information for Nanoscience

Properties near surfaces of macroscopic single crystals and of other interfaces
have been studied both theoretically and experimentally in a number of areas
over the last century. The early decades of the 20th century provided us with the
macroscopic concepts of surface chemistry, ranging from adsorption isotherms
to the dissociation of diatomic molecules and their desorption from metal
surfaces. Much of this progress is due to Langmuir, but chemists like Emmett,
Polany, Freundlich, Bodenstein, and Rideal, to name a few, have made other
major contributions.[7] From the 1960s new techniques permitted studies of
atoms and molecules on surfaces.

Over 65 techniques including photon, electron, molecule and ion scattering
as well as scanning probe methods are available today for the investigation of

composition, atomic and electronic structures, and the dynamics of their motion. Their sensitivity extends from below 1% of a monolayer up to coverages reached under high pressure conditions.[1] This makes surface science one of the main disciplines of physical chemistry. Interfacial systems can thus serve as models and provide a source of plenty and very valuable information related to nanoscale materials.

The *surface reconstruction* of silicon, which has been studied by low energy electron diffraction (LEED), may serve as an example. The driving force for such effects is the attempt of the surface to lower its energy by saturating the dangling bonds which result from the missing nearest neighbours. At the outermost surface of silicon (100) this is achieved to some extent through the formation of dimers, as illustrated in Figure 4. The relaxation resulting from this perturbation extends three to four atomic layers into the bulk.

Many metal surfaces also reconstruct. As a result of the one-sided coordination, the interatomic distances of the atoms at the topmost layer of Ir, Pt, and

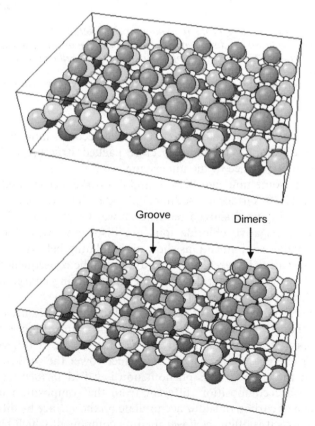

Figure 4 *Unreconstructed and reconstructed surface of a silicon (100) crystal face. Note the formation of dimers and of pronounced groves (arrows)*
(Source: http://www.chem.qmw.ac.uk/surfaces/scc/scat1_6a.htm; picture used with permission of Klaus Hermann, Fritz–Haber-Institut, Berlin.)

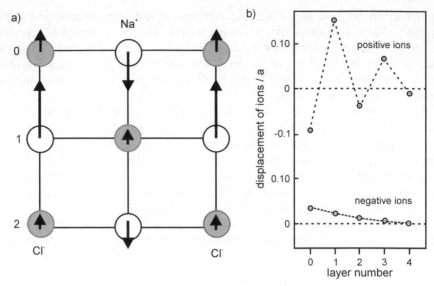

Figure 5   *Two representations of the relaxation at the (100) surface of sodium chloride.*
Arrows indicate the displacement of the ions in the 0th, 1st, and 2nd layer from
the surface. b) Displacement of ions in units of the cell constant as a function of
layer number
(Reprinted with permission from Ref. 8. Copyright American Chemical
Society)

Au (100) surfaces shrink by a few percent. It then becomes more favourable for
the top layer to adopt a hexagonally close packed structure rather than to
maintain the square lattice of the underlying layer.[1]

Already quite some time ago Benson and co-workers calculated the relaxation near the (100) surface of sodium chloride.[8] The displacements which
are displayed in Figure 5 show a tendency toward NaCl ion pair formation.
While the large negative chloride ions are displaced away from the bulk
interior, the smaller sodium cations show an oscillating behaviour. They move
in and out alternating between subsequent layers. The consequence of this is a
surface enrichment with chloride ions and an oscillating charge density as a
function of distance from the surface (compare Figure 1).

The free energy of formation of a surface is always positive since creation of a
surface from the bulk requires energy. Particles consisting of more than a single
component have, in addition to surface reconstruction, a different way of
reducing their energy: atoms or molecules which lower the surface free energy
accumulate at the surface. This phenomenon is called *surface segregation*. It
makes the surface composition different from the composition in the bulk.
Impurities such as carbon or sulfur accumulate on the surface by diffusion from
the bulk. In alloys it is often the lower melting component which has the lower
surface energy and which therefore accumulates at the surface. Thus, in a
silver–gold alloy, silver will be found in considerable excess in the first layer of
the surface.[7]

Again another option that permits stabilisation of a clean high energy surface is by adsorption of molecules which provide a lower energy surface. This often leads to *adsorbate induced restructuring*, a process which may depend significantly on the type of the adsorbate molecule. For example, a platinum (110) face obtains a quite different structure after adsorption of hydrogen, and again a different structure in presence of carbon monoxide or oxygen.[7] In the latter case the (110) planar surface restructures to a step-like surface involving (111) microfacets. This affords an increased surface area but nevertheless a lower energy. A simple explanation of this phenomenon invokes the balance between the exothermic adsorption process, giving rise to heat of adsorption, and the endothermic process of this type of surface restructuring.

Near solid–liquid interfaces it is in particular electrochemistry which has come up with a wealth of detailed and well founded models of interfacial structures. Already classic are the Helmholtz and Gouy–Chapman double layers in the electrolyte in contact with an electrode surface.[9] Closely related to this is the DLVO (Derjaguin–Landau–Verwey–Overbeek) theory of electrostatic stabilisation of colloids.[10] The precipitation of colloids is actually a common way of producing nano-clusters. In solids, gradients are of particular interest and have been investigated in detail especially near semiconductor interfaces.

More recently, defect chemistry and its effect on electrical and ionic conductivity near grain boundaries have been in the focus of solid-state electrochemistry in view of corrosion and of potential sensor applications.[11] An early and well-developed application of nano-science and technology is photography.

4 Particle Size and Microstrain

High-resolution transmission electron microscopy (HR–TEM) is a direct *ex situ* method which is used for the determination of shape and size of nanoparticles. From grain size histograms representing often only relatively small numbers of grains, average grain sizes and grain size distributions are deduced.

A more conventional method that goes back to Paul Scherrer involves the analysis of the line widths of the Bragg peaks of X-ray diffractograms.[12] He attributed the entire width to grain size effects and obtained for the full width at half maximum of a diffraction line

$$\Gamma = \frac{K\lambda}{\langle L \rangle_{\text{vol}} \cos \theta} \tag{4}$$

Here, λ is the X-ray wavelength, K the Scherrer constant which equals 0.9 for spherical particles, 2θ the scattering angle, and $\langle L \rangle_{\text{vol}}$ the volume-weighted averaged column length, *i.e.* the number of reflecting planes times their effective distance d (see Ref. 13 for a full definition). For a spherical particle $\langle L \rangle_{\text{vol}}$ equals $0.75 \langle D \rangle_{\text{vol}}$, where D is the grain diameter. The Scherrer formula is quite

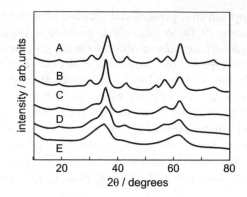

Figure 6 *X-Ray diffraction pattern of iron oxide nanoparticles ($\gamma-Fe_2O_3$) with mean diameters of 7.3 nm (A), 5.6 nm (B), 3.6 nm (C), 2.4 nm (D), and 1.9 nm (E). Note the peak broadening at lower grain size*
(Reprinted with permission from Ref. 14. Copyright American Chemical Society)

satisfactory for small grains (large broadening) in the absence of significant microstrain. An experimental example and the result of its analysis based on the Scherrer formula are given in Figure 6.

Microstrain describes the relative mean square deviation of the lattice spacing from its mean value. Based on the grain size dependence of the strain it is reasonably assumed that there is a radial strain gradient, but from X-ray diffraction only a homogeneous, volume-averaged value is obtained. Strain ε and size effect are both taken into account in the approach by Williamson and Hall.[15] On the basis of a different θ-dependence the two effects can be separated in a plot of $\Gamma \cos \theta$ against $\sin \theta$,

$$\Gamma \cos \theta = \frac{K\lambda}{\langle L \rangle_{\text{vol}}} + 4\varepsilon \sin \theta \qquad (5)$$

An alternative method that goes back to Warren and Averbach[16] utilises Fourier transformation of the reflection profiles. It allows the separation of size and strain effects and at the same time, based on the analysis of the line shapes instead of the widths only, the determination of the grain size distribution. A more advanced version combines it with a fitting procedure.[13] Using a synchrotron X-ray source it permits a reliable analysis of grains in the 5–100 nm regime, and in particular it allows *in situ* real-time studies of the grain growth kinetics with a time resolution of 5 min.[13]

It should be noted that not only the shape and width of the Bragg reflections change as a function of particle size, but also their position. In general, the peaks due to smaller particles are found at slightly higher values of θ, reflecting *lattice contraction* by a fraction of a percent or a few percent at best. The typical size-dependent behaviour of the lattice constants is given for three metals in Figure 7 together with a theoretical prediction.

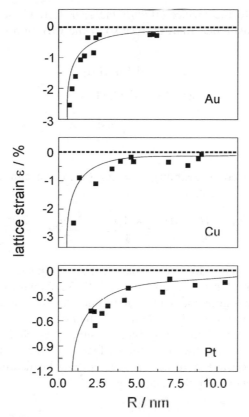

Figure 7 *Average relative lattice contraction of metal clusters as a function of size* (Reprinted with permission from Ref. 19. Copyright American Chemical Society)

The effect is a result of reduced coordination of surface atoms. Formally, it is often described by the extra pressure exerted due to surface stress. Following the derivation given by Solliard and Flueli[17] it is understood on the basis of the Laplace law which states for a liquid spherical drop of radius R that the pressure difference between the inside and the outside of its surface amounts to

$$\Delta p = \frac{2\gamma}{R} \tag{6}$$

It is immediately clear that for a surface tension γ on the order of 1 J m^{-2} the extra pressure amounts to as much as 1 GPa for a particle radius of 2 nm. We shall see later that this can stabilise the high pressure modification of a crystal structure for small values of R.

In the case of a crystalline solid in equilibrium with its vapour γ varies as a function of the crystallographic index of the crystal face, and the radius R is the distance of this face from the centre of the crystallite, measured in the direction normal to the surface (*Wulff's theorem*, Ref. 18, see also Chapter 6, Figure 13). On elastic deformation of a solid the variation of the surface free energy γA is

$$d(\gamma A) = \gamma \, dA + A \, d\gamma \tag{7}$$

and the surface stress coefficient g is defined as

$$g = \gamma + \frac{A \, d\gamma}{dA} \tag{8}$$

In terms of the compressibility $\kappa = \Delta V/(V\Delta p)$ we can write the relative change of the lattice constant

$$\frac{\Delta a}{a} = \frac{\Delta R}{R} = \frac{\Delta V}{3V} = \frac{\kappa \, \Delta p}{3} = -\frac{2\kappa g}{3R} \cong -\frac{2\kappa \gamma}{3R} \tag{9}$$

Thus, assuming that κ does not deviate from its bulk value we obtain g from the slope of a plot of Δa as a function of $1/R$. As shown in Figure 8 for two crystal faces of nano-gold one obtains good linearity. The values derived for the surface stress g amount to $3.1 \pm 7 \, \mathrm{Nm^{-1}}$ (220) and $3.2 \pm 1.0 \, \mathrm{Nm^{-1}}$ (422). This is considerably larger than the surface tension γ which is on the order of $2.0 \, \mathrm{Nm^{-1}}$. It reflects the fact that the surface is under tension and that the interatomic distances at the surface are greater than the distances characterising an infinite (111) plane of gold at equilibrium.[17]

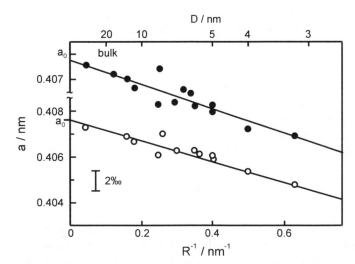

Figure 8 *Average lattice parameter in gold samples deposited on a thin carbon film as a function of the reciprocal mean particle radius R (or diameter D, respectively). The parameters are derived from the (220) diffraction peak (dots) and the (422) diffraction peaks (circles). a_0 corresponds to the bulk lattice parameter* (Reprinted from Ref. 17, with permission from Elsevier)

Careful work allows the determination of individual lattice spacings as a function of the distance from the particle centre. The example displayed in Figure 9 shows two effects: (i) the average lattice spacing decreases slightly with decreasing particle size, and (ii) there is an additional decrease near the surface of small particles.

Figure 10 displays the evolution of the microstrain with sintering time and temperature for nanocrystalline iron. As prepared by pulse electrodeposition the microstrain amounts to 0.68%. On annealing it decreases rapidly and then remains constant at limiting values between 0.3% at 663 K and 0.1% at 753 K.[13] There are thus two contributions to this microstrain, a relaxing and a nonrelaxing one. The latter apparently depends on temperature, but the authors noticed that it scales with crystallite size, since also the volume-weighted average grain diameter amounts to 19 nm as prepared, and it reaches final values of 46 nm (for tempering at 663 K), 92 nm (683 K), 234 nm (753 K)

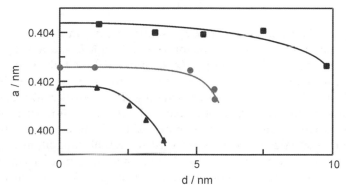

Figure 9 *Variation of the lattice parameter a with the distance d from the particle centre for MgO supported Al particles with a radius of 11.4 nm (squares), 8.0 nm (circles), and 5.2 nm (triangles)*
(Reprinted from Ref. 20, with permission from Elsevier)

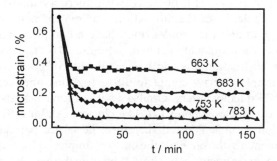

Figure 10 *Evolution of the microstrain in nanocrystalline Fe as a function of treatment temperature and time*
(Selected temperatures reprinted with permission from Ref. 13. Copyright American Chemical Society)

and 395 nm (783 K). This phenomenon reflects the fact that larger particles are thermodynamically more stable than small ones, and it is known as *Ostwald ripening* (see Chapter 6, Section 5). As a further observation, the microstrain was reported to scale with the lattice constant of the nanocrystallites.

For their application to small nanoparticles the above approaches have the principal deficit that they are based on the Bragg equation, which was derived for crystals of periodic structure and infinite size. They are not well suited to account for the strictly nonperiodic structure of quasi-crystals and of clusters with icosahedral symmetry; in particular they are also unable to describe surface restructuring in a more than phenomenological manner. A more general method uses the Debye formula[21] which gives the intensity in electron units scattered by any (crystalline or amorphous) array of atoms

$$I(k) = \sum_{i=1}^{N} \sum_{j=1}^{N} F_i F_j \frac{\sin(k \cdot r_{ij})}{k \cdot r_{ij}} \tag{10}$$

where F_i and F_j are atomic scattering factors of the i-th and the j-th atoms, $k = 4\pi\lambda^{-1} \sin\theta$ is the scattering vector at wavelength λ, and r_{ij} is the distance between these atoms. The intensity $I(k)$ is integrated over all spatial orientations of the array, so the Debye equation gives effectively a powder diffraction pattern. Thus, the equation allows calculating X-ray powder diffraction intensity profiles of any cluster, in principle without any additional assumptions about its structure, although the number of free parameters has to remain limited. Least squares procedures are used for establishing atomic configurations and taking care of grain size distributions and stacking faults in individual crystallites.[22] Nowadays, computers permit a direct (*ab initio*) calculation of diffraction profiles in hours for clusters as large as 15–20 nm in diameter, which covers the most important range of nanoparticles. Unfortunately, the computational costs scale with the square of the number of atoms in the system and thus with the sixth power of the cluster diameter.

An example of the dramatic size effect on the line width of powder diffraction profiles is given in Figure 11a, obtained with microcrystalline and nanocrystalline silicon carbide. Since the diffraction maxima for nanocrystalline materials are very broad and the profiles often have a complex shape the position of the reflection is not uniquely defined.[22] Figure 11b displays the calculated evolution of the diffraction line width as a function of grain size.

A second example displays complete diffractograms obtained with well-defined icosahedral Pd nanoclusters, stabilized with organic ligands, in comparison with bulk fcc Pd particles (Figure 12). The clusters exhibit diffraction peaks at the same position as bulk Pd, and therefore the lattice parameters calculated from the refinement are the same for the cluster samples as for bulk Pd within the accuracy of the experiment.[23] It should be noted that the organic ligands terminating the surface reduce the surface free energy and thus the pressure effect described in Figure 9.

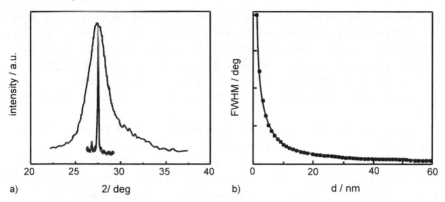

Figure 11 *(a) Two experimental X-ray powder diffraction profiles of the (111) reflection obtained with microcrystalline SiC (several μm grain size) and nanocrystalline SiC (3.9 nm grain size). (b) Full width at half maximum of SiC (111) diffraction peak as a function of grain size, obtained from 'ab initio' calculations of the diffraction pattern based on the Debye equation* (Reprinted with permission from Ref. 22)

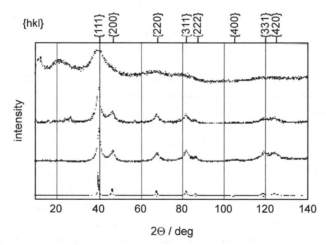

Figure 12 *X-ray diffraction pattern of Pd nanoclusters with 2 shells (Pd$_{55}$, d = 1.0 nm), 5 shells (Pd$_{561}$, 2.6 nm), 7–8 shells (Pd$_{1415}$ and Pd$_{2057}$, 3.7–4.2 nm), and Pd powder with a particle size of 50 μm Pd metal (from top to bottom), with positions of diffraction lines in fcc Pd indicated by {hkl} values* (Reprinted from Ref. 23, with permission from Elsevier)

5   Biomimetics: Nature as a Source of Inspiration for Strategies in Nanotechnology

Almost nothing in living nature is bulk. All organisms are amazing examples of highly sophisticated hierarchical composite structures which were grown spontaneously based on *supramolecular chemistry* and *self-assembling principles*. Ranging from nanometers to micrometers and millimeters they have highly variable mechanical and physicochemical properties which are determined to a large extent by the interfaces between the various components. They often combine inorganic and organic components on a nanometer scale, and they accommodate a maximum of elementary functions in the small volume of a cell.

For example, arthropode cuticles combine in different proportions chitin, proteins and calcite crystals to give tissue a rigid, flexible, opaque or translucent structure. Collagen is transparent in cornea, behaves like a low rigidity elastomer with high deformation to rupture in tendons, and associated with hydroxyapatite crystals it gives rigid and shock-resistant tissue in bone.

The structures which are built by nature are at the same time beautiful and highly reproducible and characteristic for specific organisms. How is it possible to build systematically so many different forms of diatoms, so many kinds of tiny insects, all fully functional organisms which replicate and inherit their shape and the function of each building block to their next generations so that we can categorise them and give them names? *Biomimetics* means learning from nature and copying its principles in our laboratories. It means understanding the language of shape, the driving forces of self-assembly, the strategies of template-directed synthesis, of molecular recognition, of highly selective and effective catalysis as we find it in enzymes.

Attempts to understand and mimic nature have brought disciplines back together which used to work separately and independently. When organic and inorganic materials combine at the nanoscale to give elastic or rigid elements which perhaps convert light into electrochemical energy in a living cell then we need organic and inorganic chemists, engineers, physicists, biologists and even physicians to speak one language and cooperate under the same roof. We have only just started to learn this lesson during the last two decades, but it has already resulted in amazing progress in areas such as materials science. For example, we use techniques such as molecular imprinting, structure-directed synthesis, self-assembly, take advantage of hydrophilic–hydrophobic balance and build nanoscale functional materials with tuneable physical properties and high photochemical and thermal stability and chemical inertness. We produce materials which have the strength and mechanical properties of spider silk, learning from leafs of the sacred Lotus plant we make superhydrophobic dirt-repellent surfaces. The structural analysis of dolphin skin has produced "riblets," plastic films covered with microscopic grooves which when placed on airplane wings reduce the hydrodynamic trail and economise fuel. Other materials respond to external stimuli such as solvent, pH, light, electric field, or temperature. Eventually we will design smart devices such as thin-film electronics, sensors, solar cells, biocompatible implantable fuel cells, artificial muscles, and sophisticated drug-delivery systems.

Key Points

- There are almost no properties which do not change across a surface, but none of them jumps truly discontinuously, rather they all change smoothly or in an oscillatory manner over a distance of a few atomic diameters.
- The surface-to-volume ratio scales inversely proportional to the diameter or thickness of a simple structure. This translates into the same dependence of the average coordination number of particles, and it is responsible for the $1/d$ scaling of most scalable size-dependent properties of nanomaterials.
- Atoms at surfaces, edges, kinks and corners of crystallites are of higher energy than those in the bulk. This relates to the degree of unsaturation of their bonds and is expressed suitably by their coordination number. In this respect, much can be learned about nanomaterials from surface science.
- Relaxation of structures leads to a slightly denser packing and often to pairing of atoms or ions near surfaces.
- Nanocomposite materials with amazing properties are built in nature by self-assembly. We can learn from nature and attempt to copy the principles of self-assembly to control the synthesis of advanced artificial materials.

General Reading

- G.A. Somorjai, Introduction to Surface Chemistry and Catalysis, Wiley, New York, 1994.
- C. Sanchez, H. Arribart and M.M. Giraud Guille, Biomimetism and bio-inspiration as tools for the design of innovative materials and systems, *Nature Mater.*, 2005, **4**, 277.

References

1. G.A. Somorjai, *Introduction to Surface Chemistry and Catalysis*, Wiley, New York, 1994.
2. R.L. Kautz and B.B. Schwartz, *Phys. Rev. B*, 1976, **14**, 2017.
3. J. Bardeen, *Phys. Rev. B*, 1936, **49**, 653.
4. A. Blandin, E. Daniel and J. Friedel, *Phil. Mag.*, 1959, **4**, 180.
5. D. Wolf, *Phys. Rev. Lett.*, 1992, **68**, 3315.
6. J. Jortner, *Z. Physik D*, 1992, **24**, 247.
7. G.A. Somorjai, *J. Phys. Chem. B.*, 2002, **106**, 9201.

8. G.C. Benson, P.I. Freeman and E. Dempsey, *Advances in Chemistry Series No. 33*, American Chemical Society, Washington, 1961, p. 26.
9. C.H. Hamann, A. Hamnett and W. Vielstich, *Electrochemistry*, Wiley-VCH, Weinheim, 1998.
10. J. Israelachvili, *Intermolecular and Surface Forces*, Academic Press, London, 1992.
11. J. Maier, *Festkörper – Fehler und Funktion*, Teubner Studienbücher, Stuttgart, 2000.
12. P. Scherrer, *Göttinger Nachrichten*, 1918, **2**, 98.
13. H. Natter, M. Schmelzer, M.-S. Löffler, C.E. Krill, A. Fitch and R. Hempelmann, *J. Phys. Chem. B*, 2000, **104**, 2467.
14. J.A. López-Pérez, M.A. López Quintela, J. Mira, J. Rivas and S.W. Charles, *J. Phys. Chem. B*, 1997, **101**, 8045.
15. G.K. Williamson and W.H. Hall, *Acta Metall.*, 1953, **1**, 497.
16. (a) B.E. Warren and L.E. Averbach, *J. Appl. Phys.*, 1950, **21**, 536; (b) B.E. Warren and L.E. Averbach, *J. Appl. Phys.*, 1952, **23**, 497.
17. C. Solliard and M. Flueli, *Surf. Sci.*, 1985, **156**, 487.
18. C. Herring, in *Structure and Properties of Solid Surfaces*, R. Gomer and C.S. Smith (eds), University of Chicago, Chicago, 1953, p. 5.
19. Q. Jiang, L.H. Liang and D.S. Zhao, *J. Phys. Chem. B*, 2001, **105**, 6275.
20. J. Woltersdorf, A.S. Nepijko and E. Pippel, *Surf. Sci.*, 1981, **106**, 64.
21. P. Debye, *Ann. Phys.*, 1915, **46**, 809.
22. R. Pielaszek, S. Gierlotka, S. Stelmakh, E. Grzanka and B. Palosz, *Defect and Diffusion Forum*, 2002, **208–209**, 187.
23. A. Züttl. Ch. Nützenadel, G. Schmid, D. Chartouni and L. Schlapbach, *J. Alloys Compd.*, 1999, **293–295**, 472.

CHAPTER 3

Geometric Structure, Magic Numbers and Coordination Numbers of Small Clusters

1 The Consequences of the Range of the Radial Potential Energy Function

Properties of nanoparticles strongly depend not only on their size but also on their packing structure. Experiments and computer simulations have shown that clusters of polyatomic molecules such as CCl_4, NH_3, CO_2 and hexafluorides containing as few as 100 molecules exhibit bulk-like crystalline structures. On the other hand, atomic metal or rare gas clusters up to several thousand atoms in size often exhibit local icosahedral structure which has fivefold rotational symmetry axes. This symmetry element cannot be accommodated in crystals with a long range translational repetition of unit cells. Bulk materials containing elements of fivefold symmetry do exist, but they are termed *quasi-crystalline* rather than crystalline. It has not yet been satisfactorily explained why nanoparticles built up from single atoms are icosahedral even though their bulk property is crystalline. At sufficiently small surface-to-volume ratio they will all be determined by bulk properties, but the point at which the transition happens depends on the effective range of the intermolecular forces relative to the distances between the centres of mass, on the extent of the electronic wave function that leads to the non-metal-to-metal transition, and in addition, if there is any, on the anisotropy of the repulsive part of the intermolecular potential.[1] In small particles which have a highly curved surface the surface tension translates into surface pressures which may be in the GPa range (Chapter 2, equation 6) and shift the packing structure into that of a high pressure phase of the material.

Due to the simplicity that is a consequence of the spherical shape of atoms, there have been many theoretical investigations of atomic clusters, mostly of noble gases and metals. In most cases these are based on spherical, approximate diatomic potential energy curves. A zero-order approximation is given by the hard-sphere potential

$$V(r) = \infty \quad (r \leq \sigma)$$
$$= 0 \quad (r > \sigma) \tag{1}$$

in which the potential jumps from zero to infinity as soon as the interatomic distance falls below a critical value σ. In this approximation, particles are not compressible, and they do not attract each other. It is therefore not suitable for a realistic description of most aspects of matter. Much more realistic, and therefore, often used is the *Lennard–Jones potential*

$$V(r) = 4\varepsilon\left[\left(\frac{\sigma}{r}\right)^n - \left(\frac{\sigma}{r}\right)^m\right] \tag{2}$$

where the first term in square brackets describes repulsion at small distances and the second term attraction at larger distances. Most commonly $n = 12$ and $m = 6$ are assumed (this is the Lennard–Jones-12-6 potential). σ is the interatomic distance where the potential curve crosses zero, and $|\varepsilon|$ is the binding energy at the equilibrium diatomic distance. Alternatively one uses a *Morse potential*

$$V(r) = \varepsilon[1 - \exp(-a(r - r_0))]^2 \tag{3}$$

Here, a is a stiffness or range parameter. This potential has the advantages that its shape describes chemical bonds quite realistically and that the Schrödinger equation is exactly soluble.

Of interest are the structures of the most stable clusters as a function of size and the variation of symmetry during cluster growth. The spatial constraints for packing are the same for all spherical particles. For metals, delocalisation of the conduction electrons provides an additional stability criterion that leads to a preference of certain structures, as we shall see in Section 3. This possibility does not exist for noble gas atoms. In the limit of diatomic Lennard–Jones type interaction, the size-dependent changes from one packing symmetry to another can only depend on the shape of the radial potential, that is on the relative ranges of the attractive and repulsive parts of the potential, in units of the equilibrium bond length. The point that matters is the relative contribution to the lattice energy of successive shells of atoms. For example, an atom in a body-centred cubic (bcc) lattice has 8 nearest neighbours at a distance s and 6 next-nearest neighbours at $1.155\,s$, while an atom in a face-centred cubic (fcc) lattice has 12 nearest neighbours at distance s' and 6 at $1.414\,s'$. The situation is shown in Figure 1 for a typical Lennard-Jones type potential. It is obvious that a shell at a distance of $1.6\,\sigma$ still makes an appreciable contribution so that several shells have to be taken into account. For a shorter range potential the third shell would no longer have a significant influence on the structure. We also note that there are 132 atoms within a distance of $3\,\sigma$ from a central atom for an fcc but only 128 for a bcc lattice, reflecting approximately the optimum space filling property of the former. Furthermore, many-body effects may cause deviations from the sum of diatomic contributions to energy.

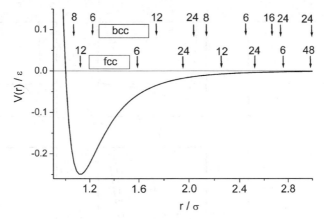

Figure 1   *Radial location of shells for bcc and fcc lattices on a Lennard–Jones-12-6 potential energy curve*

In simulations of noble gas clusters based on a Lennard–Jones potential it is common to distinguish between the different gases by varying ε and σ as element-specific parameters, but to keep the range parameters n and m fixed. In such an approximation there is no option to accommodate an independently variable range, and all elements should switch their structure at the same size, given by the same number of shells. For investigations of the effect of range it may therefore be more appropriate to base simulations on a Morse potential.

In view of their importance for nano-clusters it would be nice if we had also icosahedral structures on the graph in Figure 1. Unfortunately, icosahedra are not strictly periodic, so that the environment varies between the atoms of the different shells, and a symmetric shell-like structure is defined only for the central atom. The coordination number of the central atom is 12. Icosahedra can be regarded as slightly distorted fcc structures, but they are somewhat more compact since the interatomic distance between shells is about 5% smaller than within shells (see Section 2). As a result, the binding energy is higher by *ca.* 8% than that of equivalent sections of hexagonal close-packed (hcp) or fcc lattices.

It is quite plausible that the range of intermolecular interactions of a system is important also for its phase behaviour. This has been investigated for simple hard-sphere and Lennard–Jones potentials, and in numerical simulations also for the case of proteins.[2,3] Depending on packing density but independent of temperature, hard spheres form three phases, as shown in Figure 2a. Above a volume fraction of 54.5% all spheres are in a crystal. Below, determined by entropy, there is a narrow two-phase range comprising a fluid and a crystalline state, but below a volume fraction of 49.4% all particles are in a fluid state.

For a hard sphere potential there is no critical point below which a liquid forms. A molecular substance with a relatively long-range attractive potential of the Lennard–Jones-12-6 type leads to a typical phase diagram with solid, liquid and vapour phases as well as gas–solid, gas–liquid and liquid(fluid above the critical temperature)–solid coexistence regimes (Figure 2b). However, when

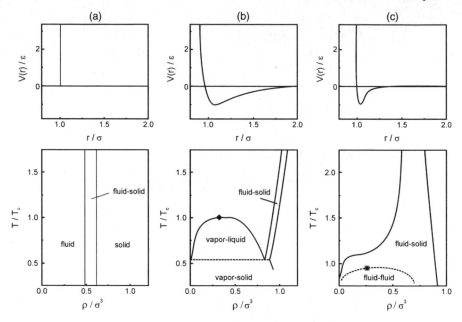

Figure 2 *(a) For a pure hard-sphere system reflected by a step-potential (upper entry) the phase diagram shows only fluid (below a freezing volume fraction ϕ_f = 0.494) and crystalline phases (above a melting volume fraction ϕ_m = 0.545). (b) Phase diagram (lower) as a function of number density ρ (units σ^3) for a Lennard–Jones-12-6 type long-range potential (upper) of molecules with effective diameter σ. (c) Phase diagram for a short-range potential that is typical for colloids with near-zero surface charge. The square indicates the dilute fluid('gas')-dense fluid('liquid') critical point. Large density fluctuations close to the fluid–fluid critical point help crystal nucleation*
(Redrawn with permission from Ref. 6. Copyright (1997) AAAS)

the range of the attraction is reduced, the fluid–fluid critical point moves towards the triple point, where the solid coexists with the dilute and dense fluid phases. If the range of attraction is made even shorter (less than 25% of the colloid diameter), only two stable phases, a fluid and a solid one, remain (Figure 2c). The reason for the difference is that the next-nearest neighbours in the second solvation shell are no longer reached so that the packing of the spheres is determined by the optimum arrangement of the nearest neighbours alone. Naturally, proteins have a large diameter relative to the range of the intermolecular interactions.

Morse potentials have a range parameter and are therefore more suitable for simulations of effects which depend on the range of a spherical potential. Moseler and Nordiek simulated clusters with pair-wise interaction represented by the Morse potential and found that the shorter the range of the potential the smaller was the difference between the melting point and the boiling point of a system.[4] The noble gases are good examples for this. The attractive part of their interaction potential is through dispersion forces and scales as r^{-6}, it is thus of short range. The melting and boiling temperatures of bulk xenon, for example, are very close to each other (161 and 165 K, respectively). Metallic clusters, having delocalised conduction electrons and therefore a relatively much larger interaction range than rare gases, are stable in the molten phase over a rather large temperature interval.

Very subtle long-range repulsive contributions to the potential can have enormous effects on a system.[3] Such types of interactions have been shown to be responsible for microphase separation, such as the formation of lamellar phases or colloids with a size of a few nanometres or even a micrometre. They are important tuneable parameters for the design of self-assembled structures.

A long range of the potential energy function normally parallels a large anharmonicity. In the anharmonic energy range the vibrational density of states is considerably larger than near the harmonic bottom of the potential. This has the consequence that the specific heat in the classical limit increases significantly beyond its value in the harmonic limit, $3k$ per atom, as given by the Dulong–Petit law. The anharmonic part of the potential is probed effectively in the liquid state, which is the reason for the fact that a system normally has a significantly higher specific heat in the liquid than in its solid state. Extremely high values were found for benzene adsorbed in the pores of NaY zeolite at a density in the order of 30% of that of bulk benzene (*i.e.* close to its critical density).[5]

Particles which remain suspended in a liquid are considered colloidal if their size is in the nanometre range, and up to several millimetres. Beyond the upper bound external forces such as gravity make them precipitate out. Examples are silver halides or gold sols, clays, silica or polymer latex particles. Colloids are of enormous importance for the preparation of many kinds of nanoparticles. The origin of the metastable state as a suspension is that they are charged and therefore repel each other. The interaction is sketched in Figure 3a. It consists of two major contributions. The van der Waals interaction, which is attractive, scales with the inverse diameter to some power ($n \approx 6$). At small distance close to physical contact it will always win and lead to coagulation or flocculation. The repulsive part of the potential is due to the interaction of the electrical double layers at the particle surfaces. It depends strongly on the electrolyte concentration in solution, which provides us with a tuning option for colloidal stability. For highly charged surfaces in a dilute electrolyte there is as a sum of the two contributions a long-range repulsion which peaks usually between 1 and 4 nm and provides a significant energy barrier towards aggregation.[7] In more concentrated electrolyte solutions there is a significant secondary minimum W (see Figure 3a), usually beyond 3 nm. The energy barrier may be too high for the particles to overcome and reach the contact minimum, so that they either sit in this minimum or remain completely dispersed in solution, rendering the colloid kinetically stable. Even higher electrolyte concentration can lead to the neutralisation of the surface charge densities so that the activation barrier between the secondary and the contact minimum disappears, which leads to coagulation. These phenomena are described quantitatively by the DLVO theory, named after work by Dejaguin, Landau, Verwey and Overbeek.[7]

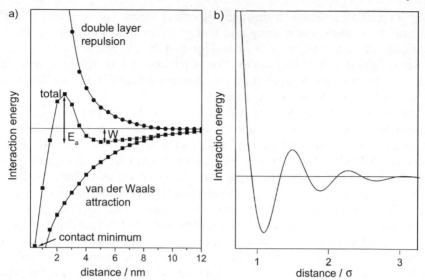

Figure 3　*(a) Schematic view of energy versus distance profiles for interactions in colloidal systems (DLVO forces). The contact minimum is slightly off scale. (b) Pair potential in liquids showing the effect of so-called 'oscillatory' forces* (Redrawn with permission from Ref. 7.)

Besides the pair potentials sketched in Figures 2 and 3a, it is instructive to also consider the potential shown in Figure 3b. In contrast to the others this does not represent a two-body interaction. Rather, it reflects the situation in a crystalline environment where atoms are found in shells in fixed distances from a central atom (compare Figure 1). The two-body pair potential may be a normal Lennard–Jones potential, but the effective multi-body potential has minima at the distance of each of the shells around a given central atom, due to the geometric congestion. Formally, it may be ascribed to 'oscillating' forces. The same situation holds also in the liquid state, except that the oscillating structure damps out more rapidly with distance. Furthermore, oscillating densities are found for particles near a flat wall, which leads to extra structure for fluids in narrow pores (see Chapter 7.6 and 7.10.2).

2　Magic Numbers by Geometric Shells Closing

The maximum coordination number of hard spheres is 12. The corresponding close packed geometry is illustrated in Figure 4.

Atoms at or near a surface have reduced coordination numbers (surface atoms in Figure 4 coordinate with five neighbouring atoms). They therefore contribute less to the overall binding energy. As a function of cluster structure the average distance of nearest and next-nearest neighbours is slightly different. Due to the element-dependent range and energetic contribution of the electronic shell structure there is a subtle system dependent energetic

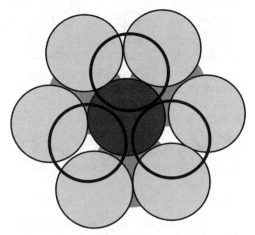

Figure 4 *Under dense packing conditions a central atom (black) is surrounded by six atoms in the same plane (light grey), three further atoms coordinate from the bottom (dark grey) and from the top (transparent circles)*

balance of cluster symmetries. Elements exhibiting electron deficient metallic bonding prefer dense packings with a large number of nearest neighbours.[8] This leads to the concept of geometrically closed structures and corresponding magic numbers. The highly symmetric clusters are of icosahedral symmetry. They have a small surface-to-volume ratio, but distances within a shell are larger than between shells by 5% so that there is an inner strain. This strain exerts a pressure on the central atom which increases with increasing cluster size. At a critical size the energy of the central atom is higher than that of an atom on the outer surface, thus it diffuses to the surface and leaves a void in the centre. Simulations with a Lennard–Jones-12-6 potential this was found to be the case for the first time at a size of 585 atoms, and all icosahedral clusters with more than 752 atoms were predicted to have a central vacancy.[9] Because of the fivefold symmetry this type cannot be realised in a crystal with periodic translational symmetry, instead it has a *shell periodicity*. Octahedral clusters correspond to fcc packing in the bulk. Since all atoms can have equal distances to all nearest neighbours they are free of strain, and together with hcp symmetry they reach the highest possible packing density of spheres ($\pi/(3\sqrt{2}) = 74\%$ for both of them, while bcc packing gives only $\sqrt{3}\pi/8 = 68\%$).

Many of the calculations impose some constraint on symmetry or bond lengths. In most cases, the highly symmetrical icosahedral structure has the lowest energy, but it should be noted that in exceptional cases disordered configurations may be favoured, as exemplified in Figure 5 for Pt_{13} clusters. In the case of Hg_{13} the optimised icosahedral geometry has a considerably higher energy than two irregular structures as well.[10] Also the charge distribution is in general not homogeneous. For example, for neutral icosahedral Pt_{13} it was calculated that the central atom was deficient by 1.6 electron charges, while for

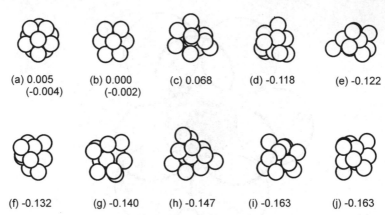

(a) 0.005   (b) 0.000   (c) 0.068   (d) -0.118   (e) -0.122
  (-0.004)   (-0.002)

(f) -0.132   (g) -0.140   (h) -0.147   (i) -0.163   (j) -0.163

Figure 5 *Different geometries for the Pt_{13} cluster and their calculated energies in eV per atom relative to the ideal cubo–octahedral (b) and icosahedral (a) clusters. The top and bottom (parentheses) values are the energies of the ideal and the Jahn–Teller distorted structures. (c–i) are local minima amorphous structures, (j) has the lowest energy*
(Redrawn from Ref. 11, with permission from the publisher)

icosahedral Al_{13} there was a surplus of 0.65 e.[11] These electron distributions reflect the occupancy of the orbitals near the Fermi level.

Clusters with a three-dimensional raspberry type structure as shown in Figure 5 are intuitively quite acceptable. It came therefore as a surprise when it was reported based on high quality density functional calculations that the ground states of Au_7^- up to probably Au_{13}^- are predicted to be planar.[12] The effect was ascribed to strong hybridisation of the atomic 5d and 6s orbitals due to relativistic effects. Moreover, it was predicted that Au_{32} has a stable hollow spherical structure akin to a fullerene molecule.[13] Again, the stability of this fascinating structure seems to be strongly rooted in relativistic effects. Another stabilising factor was ascribed to its aromaticity, as evident by ring currents which lead to strong magnetic shielding in the centre of the sphere, again resembling the findings with fullerene C_{60}. There is also evidence from ion mobility experiments that the structure of Au_N^- ($N < 13$) is planar.[14] Interestingly, while negative 55 atom clusters of neighbouring elements such as silver and copper (and also palladium and platinum) adopt a structure with clear icosahedral symmetry and therefore highly degenerate electronic states, Au_{55}^- clusters which do appear to be particularly stable have a more amorphous structure, an effect which was again ascribed to relativity.[15]

The formation of clusters is studied routinely using mass spectrometry in conjunction with a beam from an evaporation source or emerging from a nozzle. It is found that clusters of certain masses are formed with often much higher probabilities than others, and one concludes that they must be particularly stable. The numbers of atoms in such clusters of extra stability can be the same for various systems, and they were therefore termed *magic numbers*. They depend on the polyhedral geometry of the packing structure. Table 1 gives a

Table 1 *Common geometries and magic numbers of clusters*

Polyhedral geometry	Abbreviation	Magic numbers
Tetrahedron	tet	1, 4, 10, 20, 35, 56, 84, 120, 165, ...
Octahedron	oct	1, 6, 19, 44, 85, 146, 231, 344, ...
Decahedron	dec	1, 7, 23, 54, 105, 181, 287, 428, ...
Icosahedron	ico	1, 13, 55, 147, 309, 561, 923, ...
Cuboctahedron, triangular (111) surfaces	cuboct	1, 13, 55, 147, 309, 561, 923, ...
Truncated decahedron, quadrat. side planes	trdec	1, 13, 55, 147, 309, 561, 923, ...
Rhombic dodecahedron	bcc	1, 15, 65, 175, 369, 671, 1105, ...
Cuboctahedron, hexagonal (111) surfaces	troct	1, 38, 201, 586, 1289, 2406, ...
Marks truncated decahedron	mtrdec	1, 75, 147, ...

Source: From Ref. 8. Reproduced with permission from the PCCP Owner Societies. See also Ref. 16.

survey for the more commonly encountered geometries, with magic numbers calculated using the formulae given by Martin.[16]

For clusters of argon atoms or of other noble gases, which are held together by dispersion forces only, the magic numbers are 13, 55, 147, 309, ..., corresponding to completed onion-like geometric shells of icosahedral symmetry. The reason for the extra stability is that all atoms at the surface have the maximum coordination number that is possible on a surface, and therefore the average bond energy per atom is maximised. The next atom added to the surface will have a coordination number of only 3, compared to 12 in the bulk of a cluster of icosahedral symmetry.[17]

3 Magic Numbers by Electronic Shells Closing

Mass spectra of clusters of sodium atoms show a high abundance which indicates extra stability for magic numbers comprising 2, 8, 20, 40, 58, 92, 138, 196, ... (Figure 6). This reflects a more complex situation than encountered for the van der Waals clusters. We note that these are all even numbers, in contrast to the magic numbers based on geometries (Table 1). Depending on size, the stability of these clusters can be explained well by a combination of two competing concepts: geometric and electronic shells closing. We are familiar with the extra stability and inertness of closed electronic shells from the configurations of noble gas atoms. In this very same sense, sodium clusters are *pseudo-atoms* in which the delocalised 3s conduction electrons are treated as nearly free, being caged by a spherical rounded box determined by the cluster surface.[18] Solving the Schrödinger equation numerically yields discrete electronic energy levels characterised by the angular momentum

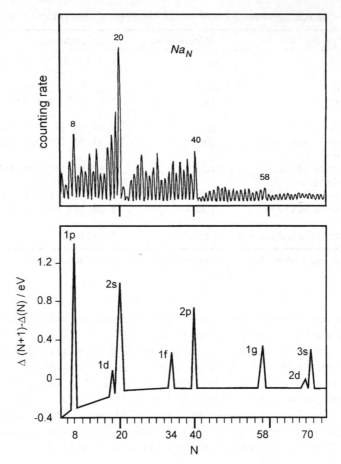

Figure 6 *Experimental mass spectrum of sodium clusters (upper) and calculated energy difference, Δ(N+1) − Δ(N) as a function of N. The labels of the peaks correspond to the closed-shell orbitals*
(Reprinted with permission from Ref. 18. Copyright (1984) by the American Physical Society)

quantum number ℓ with degeneracy $2(\ell + 1)$, including spin. The resulting energy level diagrams and cumulative occupations are shown for spherical square well, harmonic and intermediate rounded square well potentials (Figure 7a). Since the effective radius of the cluster sphere scales with $N^{1/3}$ the energy levels shift down as N increases. The extra stability of closed shell configurations then correspond to $1s^2$ ($N = 2$), $1s^2 1p^6$ ($N = 8$), $1s^2 1p^6 1d^{10} 2s^2$ ($N = 20$), *etc.* New shells start with particularly low abundance in the mass spectrum (Figure 6). It is the number of electrons which matters, Na_7^-, Na_8 and Na_9^+ are therefore to a first approximation expected to show the same shell structure, although the different size and symmetry will of course influence the situation.

The magic numbers derived from electronic shell models depend on the details of the shape of the potential function, and this varies for the different elements. For this reason, and because the energy ladders are filled to different levels, not all metals have the same packing structure in small clusters as in the bulk where the range of the potential plays a negligible role.

Density functional calculations for the real system of an icosahedral cluster $AlPb_{12}^{+}$ where 12 lead atoms are placed in a near-spherical shell around a core Al atom (Figure 7b) illustrate the similarity of the cluster molecular orbitals with the angular part of atomic orbitals (the spherical harmonics).[19] All 50 valence electrons occupy 25 molecular orbitals which are delocalised over the entire cluster. A single 1s type MO of A_g symmetry is lowest in energy, followed by a set of three degenerate 1p type (T_{1u}) orbitals which form the second energetic shell, and a fivefold degenerate set of 1d (H_g) orbitals in the third shell. This is the same sequence as predicted for the spherical square well. However, the following sevenfold degenerate 1f shell splits into a set of threefold (T_{2u}) and fourfold (G_u) degenerate orbitals. For the same reason, the ninefold degenerate 1g shell splits by 3.1 eV into a fivefold (H_g) and a fourfold (G_g) set of orbitals. These two become the highest occupied molecular orbital (HOMO) and the lowest unoccupied MO (LUMO). Furthermore, two energy levels of the second progression, 2s and 2p, are also occupied.

One can imagine that a coincidence of the closing of geometric and electronic shells leads to an even more pronounced stability than in case of closing of geometric or electronic shells alone. A popular example is that of the magic aluminium cluster anion Al_{13}^{-}, which has 40 valence electrons and forms a closed 2p shell (see Figure 7a). While the neutral Al_{13} cluster has a distorted icosahedral structure the ground state of the anion is calculated to be of almost perfect icosahedral symmetry. It is a *halogen-like superatom* and can form a molecular ion with an iodine atom, $Al_{13}I^{-}$, which has a Al_{13}–I bond dissociation energy of 2.46 eV, and the remarkable conclusion is that the negative charge remains on the aluminium cluster.[21] There have been expectations that by systematic variations it may some day be possible to draw new periodic tables of superatoms.

In analogy to the periodic table of elements with which we are all familiar we can set up periodic tables of super-atom clusters. An example is shown for sodium clusters in Figure 8, but other simple metals which contribute to the delocalised cluster orbitals with their single electron in the highest orbital, like K, Cs, or Cu, Ag and Au, would follow the same principle. The confining spherical potentials such as those shown in Figure 7a lead to an important difference in comparison with the Coulomb potential from which we derive the shell structure of normal atoms with a small positively charged nucleus: the orbital angular momentum quantum number ℓ is no longer limited to values $\leq n$ (where n is the principal quantum number). Thus, other than for normal elements, the first shell does not only consist of 1s only but has all other subshells (1p, 1d, 1f, ...). As for the elements, the energetic sequence of the subshells depends on the exact shape of the effective potential. For simple metals the sequence derived from the rounded square well potential (Figure 7a)

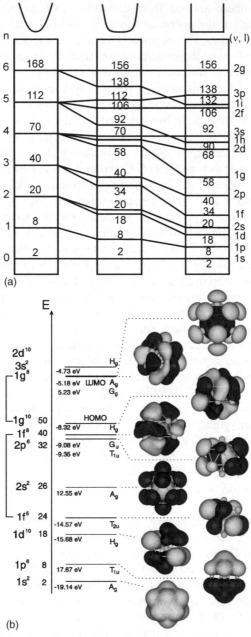

(a)

(b)

Figure 7   *(a) Energy level diagram and cumulative occupation for spherical three-*
dimensional harmonic, intermediate and square well potentials (b) Calculated
electronic orbitals (one for each set) for the core-shell icosahedral cluster
$AlPb_{12}^{+}$. *The overall cluster molecular orbitals resemble the spherical harmonics*
of atomic orbitals. On this ground, small clusters are often called 'pseudo-atoms'
((a) Reprinted with permission from Ref. 20. Copyright (1993) by the
American Physical Society).
((b) Reprinted with permission from Ref. 19. Copyright (2004) American
Physical Society).

Periodic table of elements (up to atomic number 36)

1	2	3	4	5	6	7	8	9	10	11	12	13	14	15	16	17	18
1H 1s¹																	2He 1s²
3Li [He] 2s¹	4Be [He] 2s²											5B [Be] 2p¹	6C [Be] 2p²	7N [Be] 2p³	8O [Be] 2p⁴	9F [Be] 2p⁵	10Ne [Be] 2p⁶
11Na [Ne] 3s¹	12Mg [Ne] 3s²											13Al [Mg] 3p¹	14Si [Mg] 3p²	15P [Mg] 3p³	16S [Mg] 3p⁴	17Cl [Mg] 3p⁵	18Ar [Mg] 3p⁶
19K [Ar] 4s¹	20Ca [Ar] 4s²	21Sc [Ar] 3d¹4s²	22Ti [Ar] 3d²4s²	23V [Ar] 3d³4s²	24Cr [Ar] 3d⁵4s¹	25Mn [Ar] 3d⁵4s²	26Fe [Ar] 3d⁶4s²	27Co [Ar] 3d⁷s²	28Ni [Ar] 3d⁸4s²	29Cu [Ar] 3d¹⁰4s¹	30Zn [Ar] 3d¹⁰4s²	31Ga [Zn] 4p¹	32Ge [Zn] 4p²	33As [Zn] 4p³	34Se [Zn] 4p⁴	35Br [Zn] 4p⁵	36Kr [Zn] 4p⁶

Periodic table of sodium cluster super-atoms (up to Na₃₆)

1	2	3	4	5	6	7	8	9	10	11	12	13	14	15	16	17	18
Na1 1s¹	Na2 1s²	Na3 [Na2] 1p¹	Na4 [Na2] 1p²	Na5 [Na2] 1p³	Na6 [Na2] 1p⁴	Na7 [Na2] 1p⁵	Na8 1s² 1p⁶	Na9 [Na3] 1d¹	Na10 [Na8] 1d²	Na11 [Na8] 1d³	Na12 [Na8] 1d⁴	Na13 [Na8] 1d⁵	Na14 [Na8] 1d⁶	Na15 [Na8] 1d⁷	Na16 [Na8] 1d⁸	Na17 [Na8] 1d⁹	Na18 [Na8] 1d¹⁰
Na19 [Na18] 2s¹	Na20 [Na18] 2s²	Na21 [Na20] 1f¹	Na22 [Na20] 1f²	Na23 [Na20] 1f³	Na24 [Na20] 1f⁴	Na25 [Na20] 1f⁵	Na26 [Na20] 1f⁶	Na27 [Na20] 1f⁷	Na28 [Na20] 1f⁸	Na29 [Na20] 1f⁹	Na30 [Na20] 1f¹⁰	Na31 [Na20] 1f¹¹	Na32 [Na20] 1f¹²	Na33 [Na20] 1f¹³	Na34 [Na20] 1f¹⁴	Na35 [Na34] 2p¹	Na36 [Na34] 2p²

Figure 8 *Periodic table of elements up to the atomic number 36 (upper entry) and periodic table of sodium cluster super-atoms up to Na₃₆. Colour code: s progressions (lilac), p (lime green), d (yellow) and f (pink)*

appears to be quite realistic. The subshells of Na_N clusters are thus filled by the atomic 3s electrons that are contributed by the member atoms to the delocalised cluster orbitals in the sequence 1s, 1p, 1d, 2s, 1f, 2p, ... This periodicity will of course show up in ionisation potentials and electron affinities as will be shown in Section 4 of Chapter 4.

In retrospect we note the empirical finding that the numbers of protons or neutrons that tend to be present in a stable atomic nucleus are found to be 2, 8, 20, 28, 50 and 82, which partly coincides with the above magic numbers of atoms in metal clusters. For example, tin, with 50 protons, has ten stable isotopes. The magic numbers are the numbers of protons or neutrons required to fill nuclear energy levels that have a significant gap above them. Lead has a magic number of 82 protons, which is one reason why it is the stable end-product of a radioactive decay.

The behaviour of metals like magnesium[8] or nickel[22] is quite different from that of sodium. Moreover, electronic shell models invariably have problems for large clusters since the HOMO–LUMO gap which determines the stability diminishes as $N^{-1/3}$, and details of the geometric structure become crucial.[23]

4 Cohesive Energy and Coordination Number

In a simple picture the number of bulk atoms increases with the cluster radius as r^3, that of surface atoms with r^2, while edge atoms vary linearly with r, and the number of corner atoms is constant. The cohesive energy or binding energy per atom is therefore given approximately as[8]

$$\varepsilon_{coh} \approx \varepsilon_{coh} + a_{surface}N^{-1/3} + a_{edge}N^{-2/3} + a_{comer}N^{-1} \qquad (4)$$

where the coefficients $a_i < 0$ are energy increments per atom at surface, edge or corner positions. For spherical clusters, or when corners and edges of polyhedra are neglected, a plot of ε against $N^{-1/3}$ should give a straight line which extrapolates towards ε_{bulk} for N approaching infinity. This linear behaviour was verified in a density functional theory study of magnesium clusters, as shown in Figure 9. Extrapolation yields a bulk binding energy per atom of 1.39 ± 0.01 eV for hexagonally cubic packed clusters ($N = 26, 57, 89, 103, 157$), 1.34 ± 0.04 eV for face centred cubic clusters ($N = 38, 44, 55, 116, 140, 147$) and 1.37 ± 0.03 eV for icosahedral clusters ($N = 55, 147, 309$), which is in reasonably good agreement with the experimental value of 1.51 eV in its hcp equilibrium structure. We also note that the bulk binding energy per atom is only slightly dependent on the packing geometry. In the regions of magic numbers, that is for $N = 54$–57 and for $N = 140$–153 one observes a preference for icosahedral packing. This behaviour contrasts with that of aluminium clusters where icosahedra are the least stable isomers. The different behaviour was ascribed to the larger force constant (*i.e.* shorter range potential) of aluminium which does not permit the great variation of bond lengths that is necessary for icosahedral structures.[8] In view of this subtle

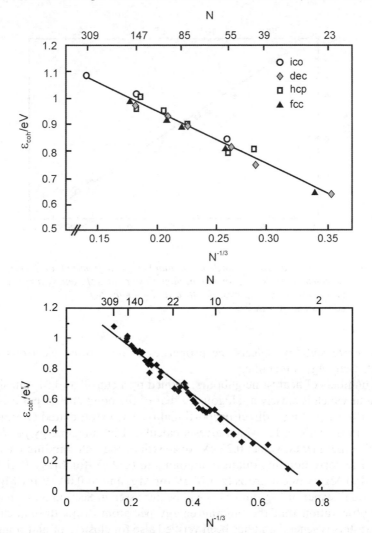

Figure 9 *Calculated cohesive energies of various size magnesium clusters as a function of*
$N^{-1/3}$ on two different scales. Upper: clusters ($N \geq 23$) with various shapes.
Lower: most stable clusters with $N \geq 2$
(Adapted from Ref. 8. Reprinted with permission from the PCCP Owner
Societies)

structure dependence it is plausible that the transition to bulk structure sometimes does not occur before a cluster size of several thousand atoms is reached.

Inspection of the lower part of Figure 9 where only the most stable structures
are displayed reveals a slight periodicity with maxima at 4, 10 and 20 atoms.
Considering that each magnesium atom contributes with two valence electrons
we recognise that these maxima correspond to 8, 20 and 40 electrons, indicating
that the extra stability is due to electronic shells closing as in the case of sodium.

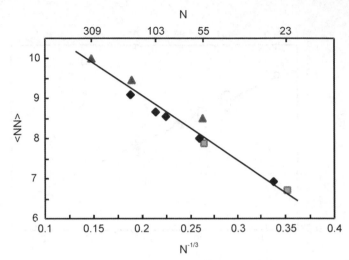

Figure 10   *Calculated average coordination number as a function of inverse radius, represented by $N^{-1/3}$, for magnesium clusters of different symmetries (triangles: icosahedra, squares: decahedra, diamonds: hcp)* (Based on data in Ref. 8).

As this extra stability reflects the properties of delocalised electrons it also suggests that Mg_4 is metallic.

The numbers of nearest neighbours, defined by a cut-off at 0.37 nm, shows a behaviour which is closely analogous to that of the bond energy per atom (see Figure 10), suggesting a direct proportionality. The extrapolated coordination number in the bulk is 12, which gives a calculated binding energy *per bond* of 0.23 eV (experimental value: 0.25 eV, one-sixth of the bulk binding energy *per atom*). The corresponding quantity amounts to 0.22 eV for Mg_{20}, 0.21 eV for tetrahedral Mg_4, while it drops to 0.14 eV for Mg_3 and to 0.09 eV for Mg_2. The special significance of Mg_3 and Mg_2 will be discussed in Section 3 of Chapter 4.

The observation that the binding energy per atom drops dramatically for small clusters is general and has been verified also for clusters of platinum[24] and of nickel.[22] As outlined above, the effect is explained to a large extent by the decrease of the average coordination number. This is also the origin of the nucleation barrier which is found for small particles in general (see Section 5 of Chapter 6).

There has been considerable effort in the investigation of the geometries and binding energies of silicon clusters. Values from various sources were collated and discussed by Bachels and Schäfer[25] (see Figure 11). For cluster sizes above $N \approx 25$ the binding energy scales as $N^{-1/3}$, as expected for spherical clusters. For $N \approx 10$–25 it is nearly constant, indicating the formation of elongated (prolate) clusters where the average coordination number does not significantly depend on the length of the aggregate. The very small clusters build compact polyhedral geometries consisting only of corners and edges. In the size range from $N = 65$ to $N = 175$ clusters with distinctly reduced binding energies of the

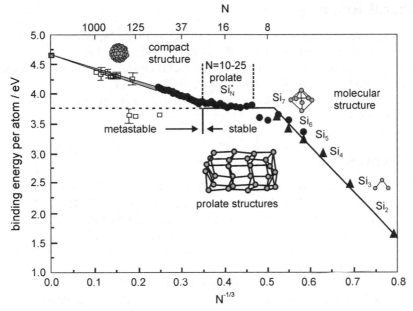

Figure 11 *Binding energies per atom of neutral silicon clusters from different experiments* (Reprinted from Ref. 25. Copyright (2000) with permission from Elsevier)

order of the one for prolate clusters were found. They were shown to be metastable and to transform to the more compact stable isomers.

Key Points

- Small particles often assume a packing structure that is different from that in the bulk, and often it lacks translational periodicity and is therefore not strictly crystalline but *quasi-crystalline.*
- The size at which the packing of nanostructures built up from atoms or spherical molecules switches to the bulk structure is related to the range of the interaction potential.
- Closing electronic shells and closing geometric shells leads to clusters of particular stability and therefore high abundance in equilibrated ensembles. The numbers of atoms in such noble gas-like closed shell configurations are called *magic numbers.*
- Clusters with a delocalised electronic structure can be regarded as *pseudo-atoms*, and the sequential filling of orbitals leads to phenomena analogous to those found along a period in the periodic table of elements.
- The average coordination number and the average binding energy per atom scale as $1/R$.

General Reading

- J. Jortner, Cluster size effects, *Z. Physik D*, 1992, **24**, 247.
- T.P. Martin, Shells of atoms, *Phys. Rep.*, 1996, **273**, 199.
- W.A. de Heer, The physics of simple metal clusters: experimental aspects and simple models, *Rev. Mod. Phys.*, 1993, **65**, 611.
- J. Israelachvili, *Intermolecular and Surface Forces*, Academic Press, London, 1992.

References

1. Y.G. Chushak and L.S. Bartell, *J. Phys. Chem. B*, 2001, **105**, 11605.
2. V.J. Anderson and H.N.W. Lekkerkerker, *Nature*, 2002, **416**, 811.
3. R.P. Sear and W.M. Gelbart, *J. Chem. Phys.*, 1999, **110**, 4582.
4. M. Moseler and J. Nordiek, *Phys. Rev. B*, 1999, **60**, 11734.
5. G. Zhao, B. Gross, H. Dilger and E. Roduner, *Phys. Chem. Chem. Phys.*, 2000, **4**, 974.
6. P.R. ten Wolde and D. Frenkel, *Science*, 1997, **277**, 1975.
7. J. Israelachvili, *Intermolecular and Surface Forces*, Academic Press, London, 1992.
8. A. Köhn, F. Weigend and R. Ahlrichs, *Phys. Chem. Chem. Phys.*, 2001, **3**, 711.
9. X. Shao, Y. Xiang and W. Cai, *Chem. Phys.*, 2004, **305**, 69.
10. B. Hartke, H.-J. Flad and M. Dolg, *Phys. Chem. Chem. Phys.*, 2001, **3**, 5121.
11. S.H. Yang, D.A. Drabold, J.B. Adams, P. Ordejón and K. Glassford, *J. Phys.: Condens. Matter*, 1997, **9**, L39.
12. H. Häkkinen, M. Moseler and U. Landmann, *Phys. Rev. Lett.*, 2002, **89**, 033401.
13. M.P. Johansson, D. Sundholm and J. Vaara, *Angew. Chem.*, 2004, **116**, 2732.
14. F. Furche, R. Ahlrichs, P. Weis, Ch. Jacob, S. Gilb, T. Bierweiler and M.M. Kappes, *J. Chem. Phys.*, 2002, **117**, 6982.
15. H. Häkkinen, M. Moseler, O. Kostko, N. Morgner, M.A. Hoffmann and B.v. Issendorf, *Phys. Rev. Lett.*, 2004, **93**, 093401.
16. T.P. Martin, *Phys. Rep.*, 1996, **273**, 199.
17. D.J. Wales and R.S. Berry, *J. Chem. Phys.*, 1995, **92**, 4473.
18. W.D. Knight, K. Clemenger, W.A. de Heer, W.A. Saunders, M.Y. Chou and M.L. Cohen, *Phys. Rev. Lett.*, 1984, **52**, 2141.
19. S. Neukermans, E. Janssens, Z.F. Chen, R.E. Silverans, P.v.R. Schleyer and P. Lievens, *Phys. Rev. Lett.*, 2004, **92**, 163401.
20. W.A. de Heer, *Rev. Mod. Phys.*, 1993, **65**, 611.
21. D.E. Bergeron, A.W. Castleman Jr., T. Morisato and S.N. Khanna, *Science*, 2004, **304**, 84.
22. V.G. Grigoryan and M. Springborg, *Phys. Chem. Chem. Phys.*, 2001, **3**, 5135.

23. J. Mansikka-Aho, J. Suhonen, S. Valkealahti, E. Hammarén and M. Manninen, in *Physics and Chemistry of Finite Size Systems: From Clusters to Crystals*, P. Jena, S.N. Khanna and B.K. Rao (eds), Kluwer Academic, Dordrecht, 1992, 157.
24. A. Sachdev, R.I. Masel and J.B. Adams, *Catal. Lett.*, 1992, **15**, 57.
25. T. Bachels and R. Schäfer, *Chem. Phys. Lett.*, 2000, **324**, 365.

CHAPTER 4
Electronic Structure

1   Discrete States Versus Band Structure

Quantum effects occur only for systems in which the valence electrons are delocalised over the entire nanoparticle. It is therefore important to understand the transition from localised to delocalised states. Atoms have a well-defined structure with discrete energy levels. For hydrogen-like one-electron atoms with nuclear charge Z the energies depend solely on the principal quantum number n according to E (eV) $= -13.6Z^2/n^2$, and the states are n^2-fold degenerate. In many-electron atoms the electron–electron interaction depends also on the magnetic quantum number ℓ ($0 \leq \ell < n$). This causes a major splitting of the states which leads to subshells with a degeneracy of $2\ell + 1$. This degeneracy is further lifted by spin-orbit interaction. We have seen in Chapter 3.3 that similar concepts apply to small clusters.

When two atoms approach each other the atomic orbitals overlap. For a proper symmetry of these orbitals, that is when the overlap integral is non–zero, the interaction leads to a splitting of the two atomic states into two molecular orbitals, a stabilised (bonding) one and a destabilised (antibonding) one. This splitting is particularly large when the two atomic orbitals are of similar energy, and it depends on the extent of overlap and thus on the interatomic distance.

It is straightforward to extend this picture to a one-dimensional (1-D) array of N equidistant identical atoms. Each set of N atomic orbitals forms a band containing N states (Figure 1). The width of a band is larger than the energy splitting of a diatom at the same interatomic distance, but it approaches an essentially constant value at a relatively low number of atoms. For large values of N this leads to a very high density of states. Low-lying states correspond to small atomic orbitals and therefore to less overlap. Thus, for a given internuclear distance α' the low energy states are nearly unperturbed, *localised atomic states* (the *core* orbitals) or very narrow bands, while the bands are wider at higher energy, and eventually they may overlap (for this reason we are often allowed in quantum chemical calculations to treat the core orbitals in a much simpler way). A metallic state is reached when a partly filled band is present or when an empty band overlaps a filled band (as for s-shells of alkali metals or d-shells of transition metals). When there is an energy gap between a filled band (*valence* band) and an empty band (*conduction band*) one has an insulator or a

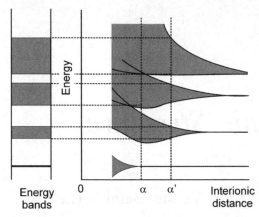

Figure 1 *Overlap of energy bands as a function of interatomic distance*

semiconductor. If the uppermost band is partly filled it is also called a conduction band. The energy of the highest filled level at absolute zero temperature is called *Fermi energy*. Conduction electrons are responsible for the finite magnetic susceptibility of bulk metals, the so-called *Pauli paramagnetism*. It is generally assumed that many aspects of band structure are realised for quite small systems of 4–20 atoms.[1] The onset of band crossing has been dubbed *Wilson transition*.

2   The Effects of Dimensionality and Symmetry in Quantum Structures

When valence electrons are delocalised their energy levels depend on the physical dimension of the system. If this dimension is in the nanoscopic range confinement of the electrons leads to quantum effects. *Quantum dots* are small self-assembled clusters of atoms such as the $AlPb_{12}^+$ ion described in Chapter 3, Figure 7b, or tailored semiconductor structures prepared by lithographic techniques. They are better regarded as molecules rather than as bulk material, and because of their discrete electronic energy spectrum they are also called *pseudo-atoms*. Since they are of nanoscopic dimension in all spatial directions ($a = b = c$ = small) they are called zero-dimensional (0-D) systems. Depending on their structure and other characteristics they may also be termed nanocrystals, or colloids if they are suspended in a liquid.

If one puts atoms or clusters into a chain one obtains *quantum wires* which have one extended spatial dimension and are therefore 1-D systems. Two-dimensional (2-D) systems are thin planar *quantum layers* or unimolecular films. The dimensionality has quite a dramatic effect on the electronic structure. If one takes an elementary concept such as that of an electron in a *D*-dimensional infinite square well, the one-electron density of states (the number of states between E and $E + dE$) per unit length, area or volume, respectively, is given within a band by[2]

$$g(E) = \frac{1}{2\pi} \left(\frac{2m}{\hbar^2}\right)^{1/2} \frac{1}{\sqrt{E}} \qquad 1-D \tag{1}$$

$$g(E) = \frac{1}{2\pi} \left(\frac{2m}{\hbar^2}\right) \qquad 2-D \tag{2}$$

$$g(E) = \left(\frac{1}{2\pi}\right)^2 \left(\frac{2m}{\hbar^2}\right)^{3/2} \sqrt{E} \qquad 3-D \tag{3}$$

A schematic representation of the energy dependence of the density of states is given in Figure 2. The 0-D system shows the expected discrete spectrum of a pseudo-atom or a molecular system, ideally with vanishing width of the states and a separation similar to that found for atoms and molecules. The 3-D case represents a macroscopic system ($a = b = c = $ large) in which the separation of the individual states becomes so small that it is no longer resolved and $g(E)$ approaches the continuous function given by Equation (3). For the 1-D system it is assumed that the extension of the remaining two dimensions is no longer negligibly small ($a >> b = c$). This causes the appearance of bands, where in each of them $g(E)$ shows an $E^{1/2}$ dependence, as given in Equation (1). The origins of the bands are given by the discrete energy levels which are determined by the narrow dimensions. Similarly, the stepwise increase of $g(E)$ of the 2-D system ($a = b >> c$) represents bands with a constant $g(E)$, as given by

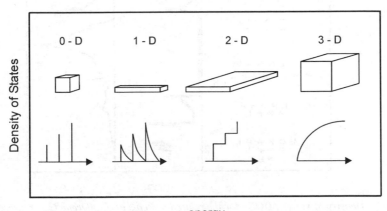

Figure 2 *Schematic representation of the effect of system dimensionality (a) Quantum dots, small clusters, colloids or nanocrystallites (0-D), ideally with discrete energy spectrum. (b) Quantum wire, chain (1-D) with non-negligible extension in the second and third dimension. (c) Quantum well, thin film or layer (2-D) with non-negligible thickness. (d) Bulk material (3-D)*
(Redrawn with permission from Ref. 3. Copyright (1996) AAAS)

Equation (2), and steps determined by the thickness of the layer. For more realistic potentials which deviate from the infinite square well model the details of the energy dependence of the density of states are governed by the exact shape of the confining potential. If in addition to the single-electron energies the electron–electron interactions and for excited states the electron–hole interactions are taken into account this will further modify the picture.[2]

Figure 3 displays a more realistic case based on the electronically delocalised all-carbon systems fullerene C_{60}, a nanotube which is a quasi 1-D system that looks like a rolled graphite sheet, and normal planar graphite. It should be noted that the density of states is non-zero in the central gap of the nanotube. This leads to electronic conductivity that can be metallic or semiconducting, depending on the way the graphite sheet has been rolled.

Many transport or optical coefficients include appropriate integrals over $g(E)$. The electronic contribution to the specific heat is proportional to the density of states at the Fermi energy, E_F. In many real systems E_F is a quantity which can be controlled either by doping or by a suitable choice of the support with which a system may exchange electrons. It can be imagined that interesting effects arise when it is tuned across a band head where the density of states changes discontinuously.

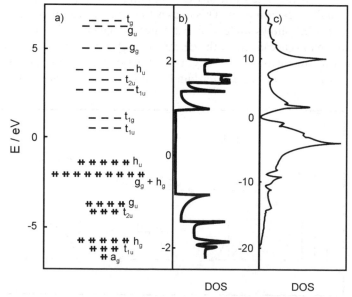

Figure 3   *Density of states (DOS) for the related systems of a fullerene C_{60}, representing the discrete states of a 0-D pseudo-atom (a), of a (10,10) carbon nanotube which is a 1-D system that shows the expected typical van Hove singularities (b), and of graphite, which is a 2-D system with a more complex behaviour than expected based on the schematic picture in Figure 2 (c)*
(a: reprinted with permission from Ref. 5. Copyright (1992) American Chemical Society; b and c: redrawn from Ref. 4, copyright (1999) with permission from Elsevier)

An important aspect in context with pseudo-atoms is *symmetry*. High symmetry can lead to a highly degenerate overall ground state of a system. In such a case, addition or removal of an electron causes a lowering of symmetry, for example by the Jahn–Teller effect. Well-known clusters of high symmetry are the fullerenes. C_{60} has icosahedral symmetry with a fivefold degenerate highest molecular orbital which is occupied by 10 electrons in the ground state (Figure 3). The lowest unoccupied molecular orbital is triply degenerate. A_3C_{60} alkali fullerides (A = alkali atom) have a symmetric, half-filled conduction band and become superconducting at low temperature.[6] C_{70} is of D_{5h} symmetry and has a doubly degenerate, fully occupied HOMO, but since C_{70} has a lower symmetry[7] there are no transitions to a superconducting state for alkali fullerides. Some of the cations and anions of C_{60} and C_{70} suffer severe Jahn–Teller distortions.[8,9]

In the infinite bulk of a solid, an atom can be in a highly symmetrical environment. The proximity of a surface breaks this symmetry. If the degeneracy of the highest occupied atomic state is lifted sufficiently, this causes a transition of the surface atoms analogous to that between a high-spin and a low-spin complex, and it may change the overall magnetic structure of a small particle relative to that of the bulk.

A prominent conventional test structure for a 2-D electron system is the thin film field-effect transistor (FET). An example that involves an organic semiconductor is shown in Figure 4. In this system we can view the semiconductor and the gate electrode as the two plates of a capacitor, for which the insulating oxide plays the role of a dielectric. A bias on the gate electrode injects charge carriers into a layer near the interface in sufficient density to make it conducting. If the organic material has a low energy LUMO it can easily accommodate extra electrons and becomes electron conducting. In contrast, electrons can be removed from materials with a high energy HOMO, making it hole conducting, as depicted in Figure 4. The morphology of the organic material in this layer is an important parameter as good orbital overlap between neighbouring molecules enhances conductivity while non-favourable geometric arrangements prevent charge hopping. Pentacene is a common small molecule that develops

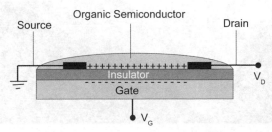

Figure 4 *Thin film field-effect transistor (FET): A negative bias V_G on the semiconductor gate injects holes into the organic semiconductor and creates a p-doped 2-D channel between source and drain which conducts electrons upon application of a drain voltage V_D. A thin insulator layer, normally SiO_2, between gate and organic semiconductor prevents a gate current. Depending on the type of material used the polarity can be the opposite*

high thin film positive hole conductivity. In single crystals it reaches amazing mobility values up to 2 cm^2 V^{-1} s^{-1}. The layer may be grown by vapour deposition or by spin or dip coating from solution. Alternatively, one may use conducting long-chain polymers such as polythiophene or polyaniline. The charge carrier density may reach close to 10^{18} cm^{-3} at a gate voltage of -30 V in a depletion layer whose thickness amounts typically to 5–20 nm.

A more recent popular 2-D structure is the interface (hetero-junction) between GaAs and Al$_x$Ga$_{1-x}$As, where x varies typically between 0.1 and 0.4. The two semiconductors are closely matched in lattice spacing, so that high-quality single crystals can be grown. In the presence of a high magnetic field perpendicular to the plane of a 2-D system the carriers are completely quantised, in that the free motion in the a–b plane condenses into Landau orbits. The density of states changes from a constant (Equation (2)) to a set of δ functions which are broadened by the effect of disorder. Evidence of this is the quantum Hall effect. As the magnetic field is swept, successive Landau levels are progressively emptied, and any quantity dependent on the density of states at the Fermi level (such as conductivity, magnetisation, electronic specific heat) exhibits strong oscillations. A review of the specific effects encountered in 2-D systems has been given by Kelly and Nicholas.[10]

A fascinating 2-D type electronic structure consisting of an elliptical fence of 36 cobalt atoms on a (111) copper surface was assembled and described by Manoharan *et al.*[11] The scanning tunnelling microscope image that was collected at 4 K resembles a corral reef or perhaps a wide open mouth of a shark. When a further cobalt atom is placed in one of the two focus points a reflected image appears at the empty focus of the ellipse. The ring with a diameter of

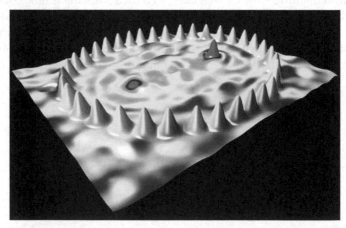

Figure 5   *Quantum mirage inside an elliptical ring of 36 Co atoms on a Cu(111) surface. A further Co atom placed at one of the two focus points inside the ellipse (purple peak) gives rise to a similar effect in the second focus point (purple spot) where no adatom exists*
(Reprinted from Ref. 11, cover picture of the issue, with permission from the publisher)

about 15 nm acts as a quantum mechanical resonator that reflects the partial waves originating from the magnetic atom at the first focus and refocuses them coherently to form a spectral image called a *quantum mirage* at the second focus. The operation of a quantum mirage is similar to how light or sound waves are focused to a single spot by optical lenses or parabolic reflectors. It demonstrates the quantum wave nature of the surface electrons of the metal. When the atom is placed elsewhere inside the ring but not at a focus point the mirage disappears. The effect is based on a Kondo resonance, a response which forms around individual magnetic moments on a metal surface (Figure 5).

3 The Nonmetal-to-Metal Transition

3.1 General Criteria

The extension of the 1-D band structure picture (Figure 1) to three dimensions is straightforward and does not lead to basic new insight. What is, however, important in the present context are the modifications of the above picture for nanoscopic systems, *i.e.* for small values of N.

- The density of states within a band is smaller by many orders of magnitude than that for macroscopic crystallites. There are only N states in a band which is derived from N non-degenerate atomic orbitals, and the density of states scales with N. For sufficiently small values of N it is possible to spectroscopically resolve discrete energy levels.
- The full width of a band may not have developed so that bands which overlap in bulk materials may be separated by a gap for small clusters. This has the consequence that nanoscopic amounts of a metal may behave as a semiconductor or an insulator, depending on the values of N and on the shape of the particle. A schematic electron level diagram corresponding to this size-induced metal-insulator transition is given in Figure 6.[12]
- Already Sir Neville Mott is cited to have expressed the view that the vexing question 'What is a metal?' has a rigorous answer 'only at $T = 0$ K, where a metal conducts, and a non-metal doesn't'.[13] On the basis of Figure 6 this view is plausible. In the presence of a small non-zero Kubo gap δ (Ref. 14) there is a continuous transition from an insulator at low temperature ($\delta >> kT$) to a semiconductor at intermediate ($\delta \approx kT$) and to a conductor at high temperatures ($\delta << kT$). Only at absolute zero temperature is there a discontinuous first-order phase transition (called the *Mott transition*) as a function of interatomic distance.
- The bulk parameters change near the surface, and the perfect periodicity of the bulk gets lost. This changes the orbital overlap of neighbouring atoms and affects the spacing of energy levels within a band. For nanoscopic amounts of material for which a large fraction of atoms is near the surface this may be noticeable in particular when it affects the electronic structure near the Fermi energy. It may mean, for example, that the interior of a cluster has metal-like properties while the surface is insulating.

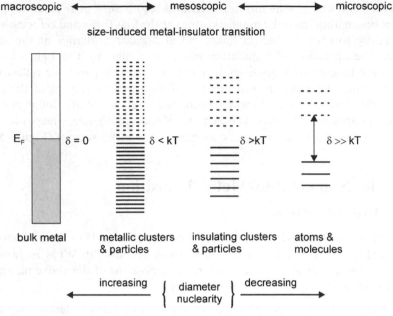

Figure 6 *Evolution of the band gap and the density of states as the number of atoms in a system increases (from right to left). δ is the so-called Kubo gap*[14] (Redrawn with permission from Ref. 12)

- In the same sense that the band structure develops and leads to a metallic state for matter under pressure, matter can lose its metallic character when it expands sufficiently, for example when the critical point is approached. This may be important for pore-confined matter where the critical point is often lowered dramatically as the pore dimension is reduced (see Chapter 7.4).
- Nanoclusters are often of icosahedral symmetry. This *quasi-crystalline state* does not have the translational periodicity of unit cells in crystalline materials. An unperturbed periodicity, however, is essential for a truly delocalised state and thus for efficient quantum mechanical transport. Bulk quasi-crystalline materials consisting of elements which count as good metals are therefore known to be bad electron conductors. This demonstrates that the existence of overlapping partly filled states is a necessary but not a sufficient condition for good conductivity. It should therefore be expected that a major increase in metallic character of nanoclusters occurs at the structural transition from icosahedral to a translationally periodic symmetry.
- Clusters of high symmetry have electronic states of high degeneracy. If these are near the Kubo gap, then breaking the symmetry for example by adding an additional atom will lift the symmetry. Some of these states may be dispersed through the Kubo gap so that adding or removing an atom may induce a nonmetal-to-metal transition.

- Low coordinated surface atoms have a high potential to form chemical bonds to molecules from the surrounding atmosphere. In this way, valence electrons which may otherwise contribute to metallic character are trapped in bonds and are no longer available as conduction electrons.
- The interplay and competition between localisation and delocalisation of electrons and the correlation of spins govern important properties such as superconductivity, heavy fermion behaviour and magnetism. Delocalisation is essential for superconductivity, localisation for magnetism.[15] This is true for bulk material, but the possibility to tune properties leaves room for the tailoring of various interesting nanoscopic materials which have no direct analogue in the bulk. However, this aspect has been little explored to date.

The essence of the discussion is that each criterion of metallic character is to some extent arbitrary, and maybe it is also not so important where we set a threshold. After all, it does not really matter whether a system is metallic or not, as long as we know its conductivity.

3.2 The Special Case of Divalent Elements

The question whether any material is a metal or a nonmetal is ultimately determined by its electronic structure. For practical work it is also a question of how metallic behaviour is determined. Several criteria can be invoked.

It is illustrative to consider and compare the case of elements with two valence electrons in an s-state, such as helium, beryllium, magnesium and mercury. The development of band structure of Mg is compared with that of Na in Figure 7. The 1s orbitals are localised close to the nuclei. Their overlap between neighbouring nuclei is small for both elements, and thus there is essentially no band formation. 2s and 2p orbitals may show limited band character, but since they are completely filled they do not contribute to metallic character. The difference between the two elements arises from the different population of the 3s orbitals. Na_2 is a diamagnetic molecule with a covalent bond, much as H_2. It has a binding energy per atom of $0.40 \, eV = 38.5 \, kJ \, mol^{-1}$. The experimental binding energy per atom of Mg_2 is only 0.025 eV, or 2.4 kJ mol^{-1}, and the molecule has a very shallow potential minimum. Both are typical for van der Waals molecules. Thus, Mg_2 behaves like the rare gas molecule He_2 of which we know well that it is only weakly bound. Na_3 is a covalently bound molecule, but it is paramagnetic, and since the unpaired electron is delocalised over all three atoms one might be tempted to call Na_3 metallic. However, the level splitting (Kubo gap) is on the order of one-half eV, so that the molecular character dominates. Mg_3, on the other hand, has all 3s molecular orbitals fully occupied, so that the bonding and antibonding character cancel approximately. Its binding energy per atom has increased to 0.14 eV, 56% of its value in the bulk. Molecular orbitals originating from s and p orbitals are still well separated so that Mg_3 has no chance of developing metallic character.

Why then is bulk magnesium a metal, and why does it develop a significant bond energy per atom ($0.25 \, eV = 24 \, kJ \, mol^{-1}$)? The origin of both effects has

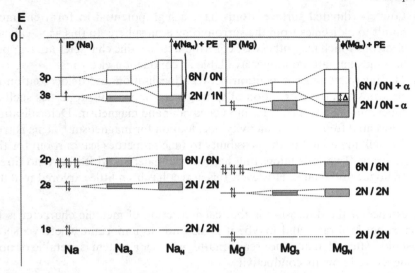

Figure 7 *Schematic representation of the evolution of band structure of sodium and magnesium in going from a single atom via a triatomic cluster to the N-atomic bulk metal. Mg₃ has a filled 3s band and cannot be metallic, whereas Na₃ has a half-filled band and can in principle be metallic, provided that the Kubo gap δ is small enough. The ionisation potential (IP) of a cluster equals the work function of the bulk metal, corrected by the size-dependent polarisation energy (PE) which disappears for the bulk (see Section 4). The total width of a band increases with the average coordination number*

to be sought in the overlap of 3s and 3p bands. The combined band is only partly filled, which makes conductivity possible. The increased binding energy may be attributed to the fact that a fraction α of antibonding 3s electrons can relax to the bottom of the 3p band which has a significantly higher density of states. This gives rise to a stabilisation of the Fermi energy by an amount indicated by Δ in Figure 7. We shall return to this scheme when we discuss ionisation potentials and work functions (Section 4).

The situation of overlapping s and p bands is equivalent to s-p hybridisation of the atomic orbitals since the extent of hybridisation or mixing increases when orbitals approach each other energetically. The degree of p-character of the valence electrons is therefore taken as another criterion for metallic character. This has been calculated for neutral Mg_N clusters, and it was found to increase from 5% in Mg_2 to 34% in Mg_{22}.[16] For bulk magnesium it approaches 50% at the Fermi level. Thus, the bonding mechanism in divalent elements changes from van der Waals over covalent to metallic.

Specifically, in order to become metallic the energy of the valence electrons in sodium has to compete with the level splitting (Kubo gap) *within* the 3s band, while a Wilson transition (overlap of 3s and 3p bands) is required in magnesium. All other alkali metals behave just like sodium, while earth alkali metals and also the group IIB elements Zn, Cd and Hg require crossing of their valence s band with the next higher p band, as detailed below for mercury. Helium, although it has also a completed s shell and thus an electron configuration analogous to that of magnesium, does not become metallic because the next empty orbital is much higher so that it is not reached in a band crossing.

3.3 Experimental Criteria of Metallic Behaviour

Various experimental tests of metallic behaviour have been devised based on the general understanding of band structures of metallic conductors.

- A suitable approach for small clusters is based on the determination of the gap between the valence and the conduction bands, and in particular its size dependence. A system that has been well investigated in this respect is mercury. This element has a ground state electron configuration with a doubly occupied s-orbital below an empty p-orbital. It is therefore of interest to study the increase of band width and progressive s–p hybridisation which goes along with band gap closure as a function of cluster size. Mass selected inner shell excitations and cohesive energy studies revealed a transition from van der Waals molecules to covalent bonding in the size range 13–20 atoms. The size dependence of the ionisation potential and quantum chemical studies of the electronic structure suggested the transition to metallic behaviour in the size range 60–100 atoms (see Section 5). More recent photoelectron spectroscopic investigations permitted a direct observation of the evolution of the band gap with size and led to a revised value of the metallic transition near 400 atoms.[17] A selection of the experimental findings is presented in Figure 8 and shows a rather impressive evolution. The electron affinity clearly increases with cluster size, and the band gap scales essentially linearly with $N^{-1/3}$ for clusters larger than about 20 atoms. The band gap closure goes along with increasing s–p hybridisation and with some bond contraction.[16] For smaller clusters there are deviations which may indicate a different electronic nature, possibly a consequence of the increasing van der Waals character of the clusters.

- A somewhat different method that is capable of providing electronic structure information of small clusters is optical absorption spectroscopy. Knickelbein[18] used the photo-dissociation of a metal cluster–rare gas van der Waals complex, for example Ni_7Ar, as a signature of photo-absorption by the metal cluster chromophore (Ni_7). Instead of distinct absorption bands as expected for small molecules, clusters of 3–7 Ni atoms were found to exhibit continuous and smooth spectra throughout the visible and near-UV regions, indicating that they absorb at every wavelength. The spectra of niobium clusters containing 7–20 atoms were likewise found to be continuous. The spectra of both types of clusters were in fact qualitatively similar to the optical response predicted by classical electrodynamics for particles of these metals containing hundreds or thousands of atoms. These observations were taken as evidence that the electronic structures of these small Ni and Nb clusters may be much closer to those of macroscopic pieces of the metals than was previously thought, in other words, it was suggested that these clusters are essentially metal-like.

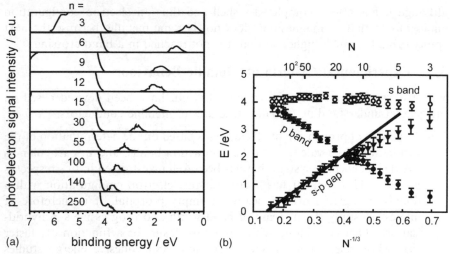

Figure 8   *(a) Photoelectron spectra of Hg_N^- in the size range of $N = 3–250$. The spectra are scaled to show a constant intensity of the single peak of the detached 6p electron. It is obvious how the band gap diminishes with cluster size. Gap closure is estimated to be complete near $N \approx 400$ atoms (selected spectra from Ref. 17). (b) Size dependence of the binding energy of the 6s HOMO (open circles) and the 6p (full circles) electrons in the photoelectron spectra of Hg_N^-. The s-p band gap (triangles) is the difference between these two values. A linear fit to the band gap extrapolates to zero near $N \approx 400$ atoms*
(Redrawn with permission from Ref. 17. Copyright (1998) by the American Physical Society)

- Similar arguments were used in the interpretation of electron paramagnetic resonance (EPR) spectra of platinum-loaded zeolites.[19] A well-resolved spectrum which revealed 12 equivalent platinum nuclei and hyperfine coupling to deuterium was assigned to possibly icosahedral $Pt_{13}D_{12}$, with negligible coupling of the central atom. The discrete nature of the spectrum was taken as evidence of the molecular rather than metallic character of the cluster. It should however be noted that the onset of changes in electron and spin delocalisation is detected by magnetic resonance methods before significant changes are observed in electron transport measurements.

- According to a theory derived by Kawabata,[20,21] metallic clusters are expected to give rise to a single EPR line of Gaussian shape due to conduction EPR. For very small spherical particles of diameter d, the peak-to-peak line width ΔH_{pp} of the derivative spectrum is given by

$$\Delta H_{pp} \cong \frac{v_F (\Delta g_\infty)^2 h\nu_e}{d\gamma_e \delta}$$

where Δg_∞ is the bulk metal conduction electron magnetic resonance g-shift from g_e, $h\nu_e$ the electron Zeeman energy, $\delta = 4\varepsilon_F/(3N)$ the average electronic level spacing (the Kubo gap energy) given in terms of the Fermi energy ε_F and the number N of energy levels in the band, v_F the Fermi velocity of the conduction electrons, which for a free electron gas is given by

$$v_F = \frac{\hbar(3\pi^2 n)^{1/3}}{m_e}$$

and n is the number density of atoms in the particle. The relation was applied to an EPR study of metal nanoparticles in mesoporous alumino-silicate molecular sieves.[22]

- There is a similar criterion, known as the *Korringa relation* that was derived for nuclear magnetic resonance (NMR) spectroscopic investigations.[23] In metals, spin-lattice relaxation proceeds via the contact interaction with polarised conduction electron spins. Electron spin fluctuations are rapid on the NMR time scale (this is no longer true for *strongly correlated* metals such as those found near a metal–nonmetal or a super-conducting transition), and Fermi–Dirac statistics applies.[24] As a result, the product of the nuclear spin-lattice relaxation time T_1 and the temperature T is a constant, given by

$$T_1 T = \frac{\hbar}{4\pi k_B K^2} \left(\frac{\gamma_e}{\gamma_n}\right)^2 B$$

where γ_e and γ_n are the electron and nuclear gyromagnetic ratios, K is the Knight shift and B a constant which is equal to unity for a metal in which many-body effects can be neglected. For bulk Pt metal, one finds experimentally a value of $B = 6$.[25] The constant Korringa product is seen as the NMR signature of the normal metallic state. It was found that it is closely obeyed for small platinum clusters with a dispersion ranging between 4 and 58%, corresponding to a diameter of ≥ 1 nm.[25] Note that T_1 becomes very long when the Knight shift vanishes.

- Direct measurement of direct current (d.c.) electrical conductivity is mainly used for macroscopic systems. In this way it was demonstrated that the conductivity of fluid oxygen increases by several orders of magnitude when it is compressed to high density. Under such conditions its valence and conduction bands broaden sufficiently so that they overlap, and the electrons become itinerant (delocalised). A limiting value of 2000 Ω^{-1} cm^{-1} was attained at 190 GPa (see Figure 9).[13,26,27] The surface pressure of small clusters is typically two orders of magnitude below this limit (see Chapter 2, Equation 6), but in cases close to the limit of metallicity it is plausible that the surface pressure can make a significant contribution to cross this limit (Figure 9). For nanoscopic amounts of matter it is more difficult to attach leads, but it has actually been possible to do so and to measure conductivity through palladium and gold nano-particles[28] and even through single atoms or molecules.[29–31] Surfactant-coated Pd particles of 15 nm diameter were contacted by two platinum tips. A linear increase of current with applied voltage as expected for

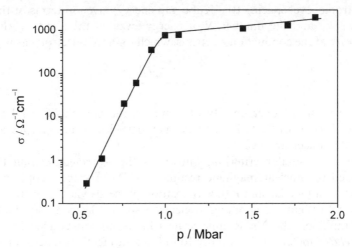

Figure 9   *Pressure dependence of the d.c. electrical conductivity of shock-compressed fluid oxygen (1 Mbar = 100 GPa)*
(Reproduced with permission from Ref. 13)

metallic (Ohmic) behaviour was found at room temperature, whereas at 4 K a characteristic stepwise behaviour was observed. This was ascribed to the so-called *Coulomb blockage*. It describes the situation when a first electron has penetrated the isolating surfactant barrier by *single electron tunnelling* and blocks further electrons from entering until it has tunnelled to the second electrode tip at a sufficiently high voltage.

- For single atom gold contacts it was found that current–voltage curves are almost Ohmic, whereas the conductance through a single atom platinum contact decreases with increasing voltage, resulting in distinct non-linear current–voltage behaviour. This difference between Au and Pt was explained by the underlying electron valence structure: Pt has an open d shell while Au has not.[32]

- A historical criterion that can be traced back to the works by Goldhammer (1913) and by Herzfeld (1927) relates to the intriguing concept of a density-induced *dielectric catastrophe* where the dielectric constant of a system would diverge at the critical density associated with a transition from the nonmetallic to the metallic state of matter.[13] In terms of the *Clausius–Mosotti* equation the dielectric constant is written as

$$\varepsilon = \frac{3 + 8\pi n\alpha}{3 - 4\pi n\alpha}$$

where n is the number density of atoms or molecules in the considered phase and α is the free atom or molecule electronic polarisability. When n approaches a critical density the denominator vanishes and ε becomes infinite, signifying a discrete phase transition to the metallic state. This allows an easy estimate of the critical density for the bulk phase. For the above example of molecular oxygen which has a polarisability α of 1.6×10^{-30} m^3 this gives a

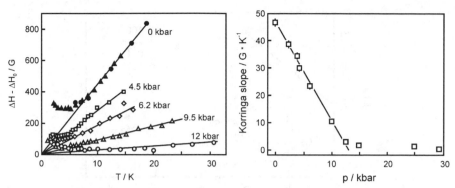

Figure 10 *Temperature dependence of the EPR line width of the Eu resonance in Yb as a function of temperature for various pressures (left), and pressure dependence of the Korringa slope (right)*
(Reproduced with permission from Ref. 33)

density of 0.25 mol cm^{-3} or 7.9 g cm^{-3}. Alkali atoms have a significantly larger polarisability ($\alpha(Na) = 24 \times 10^{-30}$ m^3), giving a critical density of 0.38 g cm^{-3}, a factor of 2.5 below the density of solid Na at standard temperature and pressure. It may be anticipated, as it is found for other phase transitions, that for finite size systems the nonmetal-to-metal transition will not be sharp.

- In this context it should be mentioned that an increase in density can also have the opposite effect. A pressure-induced metal-to-insulator transition has been studied for bulk divalent metals like Ba, Ca or Yb.[33] Here, pressure increases the hybridisation between overlapping bands, giving rise to repulsion between them, eventually opening a gap. When the Fermi level is in the middle of this gap increasing pressure leads to an insulating state. The corresponding change in electron transport can be monitored conveniently on the basis of the Korringa relation, using EPR line width measurements on doped Eu atoms. Figure 10 shows that for sufficiently high temperatures ΔH increases linearly with temperature, indicating that it is dominated by T_1 relaxation of the Eu moments due to interaction with conduction electrons in the Yb matrix. The slope which is given by the Korringa relation obviously decreases with pressure. The vanishing of the slope is associated with the vanishing of the density of states at the Fermi level, demonstrating that a hybridisation gap opens up with increasing pressure.[33]

4 Work Function, Ionisation Potential and Electron Affinity

The work function of a solid is the equivalent property to the first ionisation potential of the atom. For metals or molecular solids it is always lower than the ionisation potential of the individual atoms of which the solid is built up (see Figure 11). This is because the hole left behind on ionisation polarises the

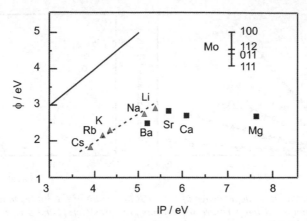

Figure 11   *Work functions φ of selected bulk metals plotted against the corresponding first atomic ionisation potentials. All entries lie well below the solid line of unit slope. Interestingly, the behaviour of alkali is qualitatively different from that of alkaline earth metals. The different entries for molybdenum demonstrate the crystal face dependence of the work function*

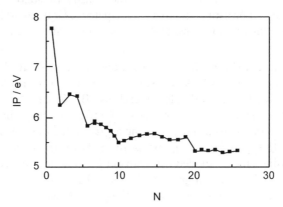

Figure 12   *Ionisation potential for iron clusters as a function of cluster size. The work function of bulk iron corresponds to 4.7 eV*
(Reproduced with permission from Ref. 36)

medium and is partially screened by the remaining electrons. The first ionisation threshold is generally observed to converge quite rapidly to the bulk work function.[34] This is demonstrated for iron clusters in Figure 12.

Theoretical models show that the size dependence of the *ionisation potential* of a cluster should follow the general relation[35]

$$\text{IP}(R, Z) = \text{IP}(\infty) + \frac{(Z + \alpha)e^2}{4\pi\varepsilon_0(R + \delta)} \tag{8}$$

where, depending on the system, IP(∞) is the bulk photoelectric threshold for molecular clusters, the top of the bulk valence band energy for semiconductor clusters and the bulk metal work function for metallic clusters. The second term on the right hand side is the Coulomb energy that is spent when the ionisation electron is removed from a spherical cluster of diameter R and initial charge Z. δ is a correction to the radius that accounts for the so-called electron spill-out. α is a fit parameter which equals $1/2$ for a macroscopic sphere, for simple metal clusters it is often close to $3/8$ but in general quite variable from 0.5 down to 0.19 for Nb.[35] α also accommodates the charge-induced stabilisation as given by Born, arising from polarisation of the medium. Since a large, polarisable cluster can stabilise the emerging charge more efficiently it is more easily ionised.

A detailed study was reported for mercury clusters. A plot against the inverse cluster diameter is given in Figure 13. We see that the ionisation potentials can be subdivided into three regions. Region III contains small clusters with fully occupied s-states from the atomic 6s orbitals. These clusters are therefore electrical insulators. Region II denotes a transition region in which overlap of 6s and 6p states sets in at approximately 13 atoms. The increase in orbital overlap reduces the ionisation potentials by about 2 eV. The larger clusters in region I have a valence electronic level structure that closely resembles the valence band structure of liquid or crystalline mercury. The ionisation potential extrapolates to the work function of 4.49 eV of bulk mercury. For small clusters, Figure 13 shows significant deviations from this simple model. They are assigned to quantum size effects, just as the oscillations in Figure 12. For a

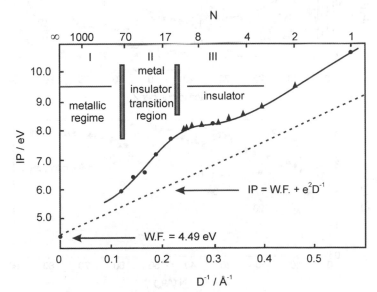

Figure 13 *Variation of the measured ionisation potential of mercury clusters as a function of cluster size. The dashed line indicates a theoretical estimate based on the classical liquid drop electrostatic model. The metal-to-insulator transition sets in at around Hg_{13},[37] a conclusion that has to be revised on the basis of more recent work (see Figure 8). W.F. = work function of the bulk solid (Reproduced with permission from Ref. 37)*

direct determination of the 6s-6p band gap we refer to Figure 8 in context with the nonmetal-to-metal transition.

The *electron affinity* (EA) characterises the reverse of the ionisation process.

$$M^{Z+} \xrightleftharpoons[\text{EA}(M^{(Z+1)+})]{\text{IP}(M^{Z+})} M^{(Z+1)+} + e^- \tag{9}$$

The extra charge that needs to be stabilised in the cluster is now on the reactant side of the process, hence EA is given by an expression analogous to Equation (8).

$$\text{EA}(R, Z+1) = \text{EA}(\infty) - \frac{(Z+\alpha)e^2}{4\pi\varepsilon_0(R+\delta)} \tag{10}$$

α is in general different for the electron affinity than for the ionisation potential. Large clusters can accommodate the additional charge more easily than small ones, electron affinity therefore increases with cluster size, as shown in Figure 14 for gold clusters, Au_N ($N = 1$–70), and compared with the familiar progressions of

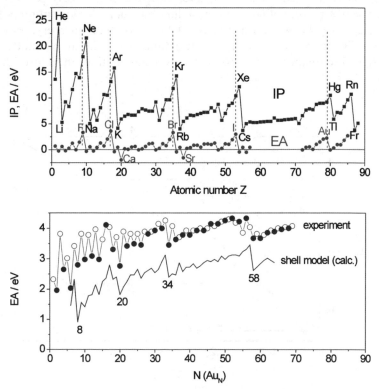

Figure 14 *Ionisation potentials and electron affinities of elements (upper tableau) in comparison with measured (circles) and on the basis of a shell model calculated (line) vertical electron affinities of Au_{1-70} as a function of cluster size (lower tableau). Even-numbered cluster sizes are marked with a full dot*
(Lower tableau reproduced with permission from Ref. 38. Copyright (1992) American Institute of Physics)

ionisation potentials and electron affinities of the elements. It is obvious that the elements and the gold *super-atoms* show analogous progressions. This illustrates nicely the electronic shell structure of clusters consisting of simple metals and supports the periodic table of super-atoms suggested in Chapter 3, Figure 8 for Na_N. In the lower tableau the clusters where the electronic shell model predicts dips when the extra electron is accommodated in a new shell are marked.[38] Odd-numbered clusters are open shell species with a singly occupied HOMO orbital. Even-numbered clusters are marked with a full dot to emphasise the alternation effect *in situ*ations of non-degenerate orbitals where each second electron is placed in an empty orbital of higher energy and therefore lower electron affinity. The size of the even–odd EA oscillations grossly reflects the spacing of the energy levels in the vicinity of the HOMO–LUMO gap. The calculated behaviour for a perturbed harmonic oscillator shell model (see Chapter 3, Figure 7a, intermediate model) reproduces well the experimental values and shows the characteristic minima at cluster sizes of $N = 8$, 20, 34 and 58. Deviations from the regular even–odd behaviour are observed near for $N = 18$ and $N = 56$ where the HOMO–LUMO gap becomes very small. Furthermore, in regions where there is a high density of states the oscillations disappear.

The calculated ionisation potentials and electron affinities for magnesium clusters are displayed in Figure 15. They reflect nicely the symmetry suggested by the two Equations (8) and (10). It should be noted that for bulk metals the ionisation potential $IP(\infty)$ and the electron affinity $EA(\infty)$ are equal since the extra or missing charge at the Fermi level delocalises over the entire system so that the energy of the upper edge of the conduction band is not significantly affected.

Interesting linear relations were reported about the cube root of the static dipole polarisability against the inverse first ionisation potential (Figure 16),

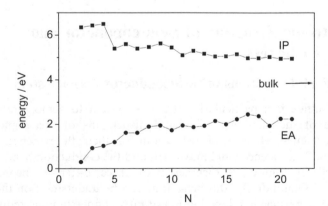

Figure 15 *Calculated vertical ionisation potentials and electron affinities (calculated as vertical detachment energy of the anion) of neutral Mg_{2-21} as a function of cluster size. The arrow indicates the common value of 3.66 eV for IP and EA of the bulk metal*
(Reproduced with permission from Ref. 16. Copyright (2002) American Chemical Society)

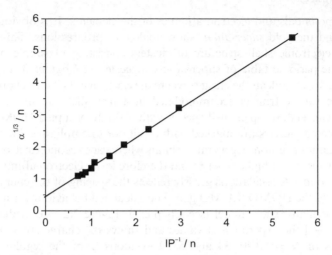

Figure 16 *Cube root of the experimental polarisability α against the calculated inverse first ionisation potential, both divided by the number of atoms in the cluster, for sodium clusters (properties in atomic units)*
(Reproduced with permission from Ref. 39. Copyright (2004) American Chemical Society)

and against the softness (not shown) of sodium clusters of up to 10 atoms.[39] The polarisability, which scales approximately as R^3, is a measure of the distortion of the electron density under the effect of an external static electric field. According to Born the charge left behind after ionisation is stabilised by polarising the cluster, and obviously the higher this stabilisation the lower the ionisation potential (compare Equation (8)).

5 Electronic Structure of Semiconductor and Metal Clusters

5.1 Optical Transitions in Semiconductor Nanoclusters

Cadmium sulfide is a particularly attractive system to demonstrate the size dependence of the band gap. Colloidal dispersions of this semiconductor display spectacular changes of the colour of their fluorescence. The size dependence of the valence band maximum and the conduction band minimum is displayed in Figure 17 together with three different theoretical approaches.[40,41] Qualitatively, this behaviour can be understood on the basis of the model of a particle in a box, but quantitative understanding requires more sophisticated treatments. It is seen that the valence and conduction band shift contribute about equally to the total band gap opening of about 2 eV with respect to CdS bulk as the particle size decreases to 1 nm. Taking a finite height of the potential well as 2.7 eV into account the experimental results can be reproduced within an effective mass approximation model.[40] It should be

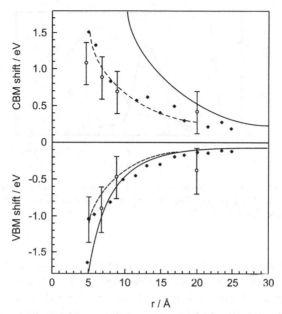

Figure 17 *Size dependence of the band gap of CdS nanocrystals. Shown are the experimental valence band maximum (VBM) and the conduction band minimum (CBM) together with calculations based on infinite (full line) and finite (broken line) potential wells and tight binding calculations (diamonds)* (Reprinted from Ref. 40. Copyright (1999) with permission from Elsevier)

stressed that the band gap does not disappear for large particles and that even the bulk material is a semiconductor.

One of the most attractive and fascinating aspects of size-dependent phenomena is that the colour of luminescent light of CdTe can be tuned all the way from green to red just by adjusting the particle size from 2 to 5 nm. The microcrystalline semiconductors CdS and CdSe have been intensively investigated in the form of colloidal suspensions in liquids, or dispersed in a glass or rock salt matrix. They have large size-dependent energy gaps, and in glassy matrix they are widely used as cut-off colour filters for the visible part of the optical spectrum. The band gap can be tuned between 4.5 and 2.5 eV in CdS, and between 2.4 and 1.7 eV in CdSe, as the size is varied from the molecular regime to the macroscopic crystal, and the radiative lifetime for the lowest allowed optical excitation ranges from tens of picoseconds to several nanoseconds.[3]

Experimental evidence for quantum size effects of *excitons* confined in all three dimensions was obtained by Ekimov and Onushchenko[41] for CuCl dispersed in silicate glass. As shown in Figure 18, the band shifts to higher energy for smaller crystallites are quite obvious. Note also the enhanced volume-normalised oscillator strength as the crystallite radius is reduced. This arises because the oscillator strength becomes concentrated over sharp electron–hole transitions, rather than being distributed over a continuum of states as for bulk semiconductors.

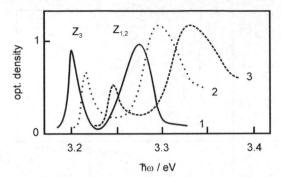

Figure 18   *Shift of the exciton peak Z_3 and the degenerate peak $Z_{1,2}$ observed at 4.2 K with CuCl microcrystallites having an average radius of 3.1 nm (1), 2.9 nm (2) and 2.0 nm (3).[2,42]*

The observation led to a theoretical treatment by Efros and Efros[43] which served as a basis for all subsequent refined treatments. The theory assumed a spherical potential of infinite height and a size defined by the radius of the microcrystallite, and the effective mass approximation was applied for the excited electron and the hole which is left behind. The main energy terms are the electron–hole interaction energy (the Coulomb term) and the confinement energy of electron and hole, a kinetic energy term.[2] Denoting the average crystallite radius as R, three cases are distinguished:[2]

- $R >> a_B$ ($R >> a_e$ and $R >> a_h$), where a_B is the Bohr radius of the exciton in the bulk semiconductor, and $a_B = a_e + a_h$, where a_e and a_h are the electron and hole Bohr radii, respectively. This is the region of weak confinement where the Coulomb term dominates, and there is a size quantisation of the translational motion of the exciton as an entity. The shift of the energy levels scales as R^{-2}, and the energy of the ground state exciton is given by

$$\Delta E \approx \frac{h^2}{8MR^2} \quad \text{(weak confinement)} \tag{11}$$

where the exciton mass $M = m_e^* + m_h^*$ is given by the sum of the electron and hole effective masses (Figure 19).

- $R << a_e$ and $R << a_h$ is the region of strong confinement. The Coulomb term is small and can be neglected or treated as a perturbation (there is no real exciton formation). Electron and hole can be viewed as independent confined particles with separate size quantisation. The lowest energy transition occurs then at

$$\Delta E \approx \frac{h^2}{8\mu R^2} \quad \text{(strong confinement)} \tag{12}$$

The exciton mass M is replaced by the reduced exciton mass $\mu = m_e^* m_h^* / (m_e^* + m_h^*)$. Strong confinement leads to a ladder of discrete energy levels, as in a molecular system, rather than energy bands.

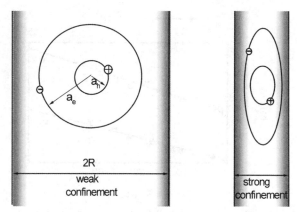

Figure 19 *Schematic representation of weak and strong confinement of an electron–hole pair in a semiconductor nanocrystal of diameter 2R. In a thin plate as shown on the right, or in a sufficiently small crystallite, the grain boundaries limit the natural extension of the exciton*

- The third case is for the usual situation that is encountered in very small crystallites with $R << a_e$ but $R >> a_h$. Since the effective mass of the hole is much larger than that of the electron, μ can now be replaced by m_e^*. The electron is quantised, and the hole interacts with it through the Coulomb potential, so that we obtain

$$\Delta E \approx \frac{h^2}{8m_e^* R^2} \quad \text{(electron confinement)} \tag{13}$$

In all these cases the Bohr radii are given in SI units by

$$a_e = \frac{h^2 \varepsilon_0 \varepsilon_2}{4\pi m_e^* e^2}, \qquad a_h = \frac{h^2 \varepsilon_0 \varepsilon_2}{4\pi m_h^* e^2}, \qquad a_B = \frac{h^2 \varepsilon_0 \varepsilon_2}{4\pi \mu e^2}$$

ε_2 is the background dielectric constant of the semiconductor material. In contrast to many size-dependent properties which we have encountered so far the exciton excitation energy scales with the inverse square radius, not with $1/R$. In Equations (11–13) we recognise the close analogy to the results of a particle in an infinite square well.

For the lowest energy transition Z_3 the excitation energy can be written as

$$\Delta E(Z_3) = E_g - E_b + \frac{0.67 h^2}{8MR^2} \tag{15}$$

where E_g is the gap energy and E_b the exciton binding energy in the bulk ($R \rightarrow \infty$). A similar expression holds for the degenerate transitions $Z_{1,2}$.[42] The linear behaviour as a function of R^{-2} is demonstrated in Figure 20 for CuCl. The offset of the two straight lines at infinite radius represents a spin-orbit splitting. From the slopes one obtains $M = (1.9 \pm 0.2)m_e$ (for the Z_3 exciton),

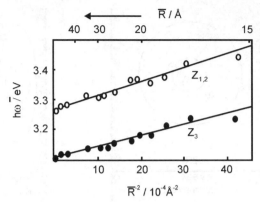

Figure 20 *Energies of two excitons for CuCl microcrystals as a function of* R^{-2}.[42]

the 'light' exciton mass $M_l = (1.5 \pm 0.2)m_e$ and the 'heavy' exciton mass $M_h = (2.6 \pm 0.2)m_e$ (for $Z_{1,2}$). It is emphasised that the effective mass approximation describes the experimental results for crystallite radii down to about 1.5 nm.

For a semiconductor the difference between the bulk and a cluster is understood basically in the same way as for metals and insulators (compare Figure 6). As illustrated in Figure 21, the gap is expanded in the cluster. This affects also the energies of the defect states, a question that was addressed by Chestnoy et al.[44] It is normally considered that clusters of about 200 member atoms have already the same unit cell and bond lengths as the bulk material. Nevertheless, clusters of II-VI semiconductors such as ZnS or CdS should contain at least 10^4 atoms before bulk behaviour is fully developed.[2]

A major difference arises from the fact that a crystalline bulk semiconductor has relatively few well-defined point defects such as missing or oxidised atoms, while a small nanoparticle is often dominated by its irregular surface which is composed of kinks, edges and corners, each with atoms of slightly different chemical potential. At the surface of a pure tetragonal semiconductor such as silicon, substantial reconstructions in the atomic positions occur to partially saturate the dangling bonds. It is found that these *surface states* invariably lead to energy levels within the forbidden gap of the bulk solid (Figure 21).[3] These surface states can trap electrons or holes and influence the electronic and optical properties of a material considerably.

The surface states are passivated when the surface atoms can form a chemical bond to another material with a much higher band gap. Ideally, such termination prevents any reconstruction of the surface, leaving no strain and leading to a much more sudden jump of the potential which confines electrons and holes, so that the situation is described well by the particle in a box model. In colloidal preparations of CdS the surface is often terminated by thiophenol or mercaptoacetic acid to provide solubility. However, organic ligands are labile and in dynamic equilibrium with the surrounding medium. This often leads to slow deterioration of the photophysical properties due to surface oxidation.

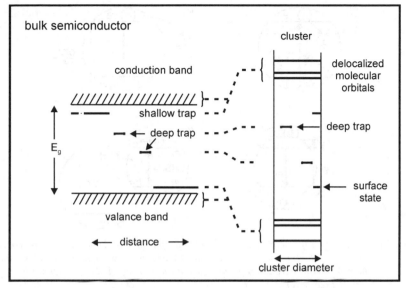

Figure 21   *Schematic energy level diagram for the bulk semiconductor and the cluster or microcrystallite*
(Reproduced with permission from Ref. 44, Copyright (1986) American Chemical Society)

5.2   Photochemical and Photophysical Processes of Semiconductor Nanoparticles

The photophysical and photochemical processes in semiconductor nanoparticles are quite analogous to those of molecules which are well known. It has become customary to describe these processes by drawing the valence and conduction band levels inside circles, representing the spherical clusters, as shown in Figure 22.[45,46] The basic process upon absorption of light is that the electron e in the conduction band and the hole h that is left behind in the valence band both relax to trapped states e_t and h_t, normally at the surface, before they recombine under emission of red-shifted fluorescence (Figure 22a). Alternatively, if the particle is in contact with electron acceptor molecules A the electron is transferred under formation of A^-, and contact with suitable hole acceptor (electron donor) molecules D gives D^+ (Figure 22b).

Of considerable interest are concentric *core-shell heterostructures* consisting of two or more materials with different energy levels. An example is given in Figure 22c where the shell material has a larger band gap and both energy levels are displaced symmetrically about the gap centre of the core material. In this case, when the shell is excited, both electron and hole relax to the core particle where they are protected from reaction with quenchers in the outer environment of the particle. Thus, such a structure increases the fluorescence yield and lifetime. Inverting this shell structure will lead to segregation of the defects at

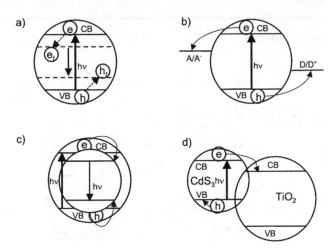

Figure 22 *Schematic representation of typical photophysical and photochemical processes in semiconductor nanoparticles. The circles represent the nanoparticles, and the horizontal lines are the edge energies of the valence (VB) and conduction (CB) bands (see text)*

the surface. A structure with a surface well was realised by Weller and co-workers[47] by coating a CdS crystal with a monolayer of HgS, followed by a final layer of CdS. In principle, core-shell systems have the advantage that the number of electrons in a system can be varied without changing the number of atoms; the closing of geometric and electronic shells can therefore be tuned independently.

When the lattice mismatch between the core and the shell structure is too large this leads to lattice strain, inducing misfit dislocations at the interface, which serve as traps and affect negatively both the photoluminescence efficiency and the stability of the nanocrystals. In such a case it is advisable to introduce a thin layer of a third semiconductor, sandwiched between the core and the outer shell, with intermediate lattice parameters. In the CdSe, CdS, ZnSe, ZnS family with wurzite structure the bulk band gap varies from 1.8 to 3.8 eV and scales approximately linearly with lattice parameters. ZnS is a non-toxic, chemically stable material which is well suited as a robust outer shell; moreover, its high band gap provides the necessary potential to confine both electrons and holes inside the core. On this basis, CdSe/CdS/ZnS and CdSe/ZnSe/ZnS core-shell-shell nanocrystals were synthesised successfully, providing photoluminescence efficiency and photostability which exceed those of CdSe/ZnS nanocrystals.[48]

Another interesting case which works on the same principle as the core-shell systems is that of materials which consist of a mixture of two different nanoparticles in contact (Figure 22d). If CdS is excited in contact with TiO_2 the conduction band electron will be transferred to TiO_2 which has its conduction band at a lower level by 0.5 eV (bulk value), while it is unfavourable for the hole to be transferred. Thus, this will lead to fast charge separation, a process which is essential in solar cells. Similarly, a $CdS/TiO_2/CdS$ three-component core-shell system is a quantum well structure in which the electrons tend to get trapped in a TiO_2 shell of spherical symmetry. Due to quantum size effects, the conduction band of smaller TiO_2 particles may be above that of CdS, so that electron transfer is found to depend on size.[49]

Absorption and luminescence spectra of three colloidal dispersions of monodisperse CdSe nanoparticles are shown in Figure 23. It is obvious that the absorption edge and the first absorption maximum shift to higher energy as the particle size decreases. The most striking difference, however, is the width of the emission peak. The luminescence of the 3.2 nm particles is dominated by narrow, near-band-edge or shallow trap luminescence, while deep-trap emission dominates in the 1.6 nm particles. The authors conclude that the particle surface is strongly non-spherical, and that the luminescence originates from traps which are much lower in energy than the absorbing state.[50] As the particle size decreases, surface interactions dominate the luminescence properties. Intermediate size particles show behaviour with both emissions occurring in parallel.

While the hole is heavy and entirely confined inside a crystallite the electron may spend some time outside the surface. The presence of adsorbates or contaminants at the surface, and the environment in general, is therefore

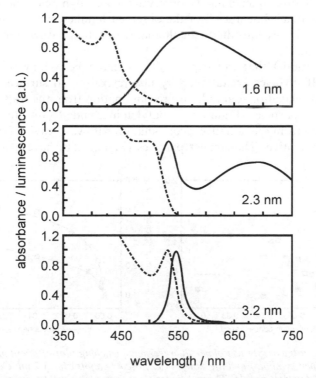

Figure 23 *Absorption (dotted lines) and luminescence spectra (full lines) of three monodisperse CdSe nanoparticle dispersions in toluene with diameters as given in the figure*
(Reproduced with permission from Ref. 50. Copyright (2001) American Chemical Society)

important for the optical behaviour of a cluster. This has been investigated in a number of studies. For example, adsorbed amines which act as hole acceptors influence the optoelectronic properties. This is shown in Figure 24 which displays Stern–Volmer quenching plots as a function of butylamine concentration for 3.2 nm (top) and for 1.6 nm (bottom) CdSe particles. A non-linear quenching of the luminescence quantum yield is observed. The luminescence lifetime is largely unaffected for the larger particles (top) but much less so for the smaller particles (bottom).

The fact that the decay lifetime is not quenched in a similar way as the quantum yield indicates that dynamic electron transfer is not the origin of the quenching effect; rather, the observations may be ascribed to static binding of the butylamine to trapping sites on the surface.[50] Since these trapping sites appear to be involved in the radiative recombination process, this removes the emitting centres from the recombination and thus prevents luminescence. Nanoparticles to which butylamine has adsorbed do not emit light at all. The other nanoparticles which are free of butylamine luminesce as if no butylamine had been added.

The smaller particles are quenched much more efficiently than the larger ones, and lifetime quenching is observed. Thus, dynamic quenching now accounts for much but not all of the observed luminescence quenching. The amine has obviously introduced an alternate relaxation pathway to the excited charge carriers.

A simple model that can explain the observed behaviour is depicted in Figure 25. It involves excitation, very fast relaxation to surface trap states, radiationless relaxation and radiative recombination from the trap. In the case of the 3.2 nm particles (Figure 25a), butylamine binds to the trap states and removes them from the radiative recombination pathway, leaving non-radiative relaxation operative. The smaller particles (Figure 25b) have a multitude of

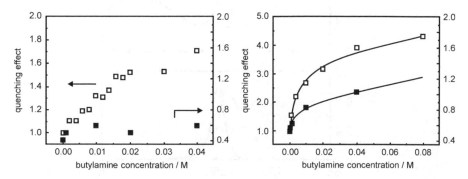

Figure 24   *Non-linear Stern–Volmer fluorescence quenching plot of quantum yield (I_0/I, open symbols) and life time (τ_0/τ, closed symbols) for 3.2 nm CdSe nanoparticles (left panel). The non-linearity observed with the 1.6 nm particles (right panel) illustrates that the quenching of the emission can be at least partly attributed to dynamic electron transfer*
(Reproduced with permission from Ref. 50. Copyright (2001) American Chemical Society)

a)

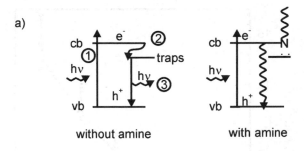

without amine   with amine

b)

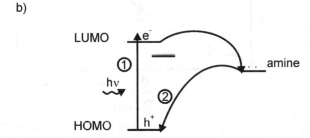

Figure 25 *Model of the luminescence process involving excitation (1) from the valence band (vb) to the conduction band (cb), very fast relaxation to surface trap sites (2), and radiative recombination from the trap (3). Behaviour of (a) 3.2 nm and of (b) 1.2 nm particles*
(Redrawn with permission from Ref. 50. Copyright (2001) American Chemical Society)

surface irregularities (deep-trap states) which dominate the interactions of the particle with the surrounding medium. Also, the gap has widened as a consequence of the smaller size. Electron transfer between the amine and the particle is energetically favourable because of the wider band gap and the introduction of the deep-trap emission pathway (2).

The absorption cross section of CdSe nanocrystal quantum dots was measured in function of size and compared with theory by Leatherdale *et al.*[51] When the volume fraction of particles is much less than unity, the absorption cross section far from any strong resonances and far from the band edge scales with the isotropic polarisability of the particle, which in turn is proportional to the particle volume and to the refractive index difference between the particle inside and outside. Furthermore, it is inversely proportional to the exciting wavelength. Figure 26 shows that theory can account well for the experimental findings.

5.3 Optical Properties of Metal Nanoclusters

Comparison of Figure 23 with Figure 27 demonstrates the fact that absorption spectra of metallic nanoparticles may resemble those of semiconductor clusters, even though the origin of the resonant absorption band may be quite different. Absorption due to single-electron inter-band transitions of core electrons often does play a role. However, the presence of conduction electrons which can be

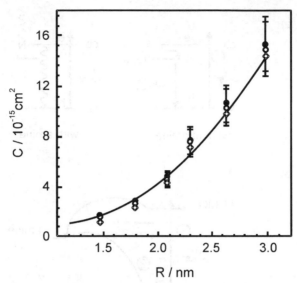

Figure 26 *Observed and theoretical absorption cross section as a function of size for CdSe nanocrystals dispersed in hexane. The symbols represent experimental values at 350 nm from different determinations with and without corrections for absolute reaction yields, and the solid line is the theoretical curve based on literature parameters for CdSe*
(Reproduced with permission from Ref. 51. Copyright (2002) American Chemical Society)

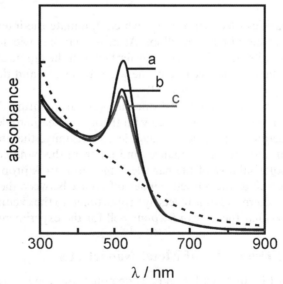

Figure 27 *Optical absorption of colloidal gold nanoparticles with average diameters of 10 nm (a), 7.5 nm (b), 6.0 nm (c) and after laser-induced size reduction (dotted line)*
(Reproduced with permission from Ref. 54. Copyright (2002) American Chemical Society)

excited collectively complicates the situation considerably. A detailed account of the complex process involved has been given by Kreibig and Volmer.[52] Mie[53] had already stated in 1908 that 'because gold atoms surely differ in their optical properties from small gold spheres it would probably be interesting to study the absorption of solutions with the smallest submicroscopical particles; thus in a way, one could investigate by optical means how gold particles are composed of atoms'. Initiated by this idea he developed an exact electrodynamic calculation (nowadays called the *Mie theory*) based on Maxwell's equations to account for the frequency-dependent response of spherical metal clusters to an electric field. This response is governed by the real (polarisation) and the imaginary (energy dissipative) parts of the dielectric function, $\varepsilon(\omega) = \varepsilon_1(\omega) + i\varepsilon_2(\omega)$. For metals in the limit of a static electric field ($\omega \to 0$), ε_1 diverges to $-\infty$. The quasi-free conduction electrons can undergo *collective excitations* which result in standing waves. They obey certain resonance conditions and are therefore called *plasmon resonances*. Band positions, heights and widths depend sensitively on the dielectric function, and indirectly, via the size dependence of $\varepsilon_{1,2}(\omega)$, also on the particle radius.

Several theoretical models predict a correction to ε_2 that is proportional to $1/R$ and thus reflects the competition between surface and volume of the cluster.[52] ε_1 and ε_2 are related via the Kramers–Kronig relation. The most prominent feature of the optical response of large clusters (as long as $R << \lambda$) of metals where the free electron approximation holds well (such as alkali and noble metals, and aluminium) is the dipolar surface plasmon. Its position is defined by the condition $\varepsilon_1 = -2\varepsilon_m$ (ε_m is the dielectric constant of the medium surrounding the cluster, *i.e.* $\varepsilon_m = 1$ for vacuum). Clusters above a critical radius which amounts to typically 5 nm should exhibit a red-shift of the surface plasmon resonance which increases with R due to phase retardation of the electromagnetic wave in the metal. Smaller clusters can exhibit red-shifts as well, but with decreasing R, due to spill-over of conduction electrons at the cluster surface, and this effect may be offset by quantum size effects. As discussed in Section 3, very small clusters lose their metallic character, and collective excitations are suppressed or they obtain increasingly single-electron character so that the term plasmon resonance loses its meaning.

Figure 27 shows an example of the size dependence for colloidal gold particles. We see that in this case it is especially the width and not the position of the resonance that is affected by the cluster size.

An important parameter is the dielectric constant of the surrounding medium. Due to the electrodynamic boundary conditions at the cluster surface or interface the position of the plasmon resonance depends on the refractive index of the embedding medium. The technique which makes use of this effect is called *immersion spectroscopy*. By selecting a particular matrix material with $1 \le \varepsilon_m \le 5$ a resonance can be markedly enhanced by shifting it away from inter-band transitions. This permits studying core-shell structures, as for example of surface-oxidised or surface-melted metals. The resonance of a solid gold particle is strongly damped and thus smeared out when it is embedded in a liquid Au shell.

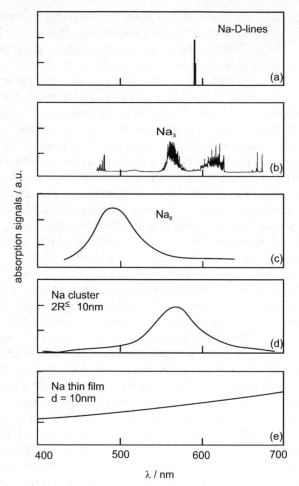

Figure 28 *Optical absorption of sodium as a single atom (a), a triatomic cluster (b), a free Na$_8$ cluster in a molecular beam (c), a large cluster of the order of 10^5–10^6 atoms and a diameter of around 10 nm in an NaCl matrix (d) and a thin film (e)*
(Reproduced from Ref. 52. Copyright (1995) with permission from Springer)

Figure 28 demonstrates the striking development of electronic spectra of elemental sodium as one goes from the single atom to free clusters of increasing size, and finally to a thin film. In the given wave length range, the atom shows only the familiar yellow D-line doublet absorption near 589 nm. The triatomic cluster exhibits a typical molecular character with absorption bands with vibrational and rotational fine structure. In contrast, an eight-atom cluster has a smooth spectrum with a single band, indicating that the molecular character has been lost, possibly indicating the onset of metallic behaviour. On further increase to about 10 nm, this band is red-shifted, and a thin film of similar thickness displays a monotonic increase over the given wave length range that is typical for free electron behaviour in a bulk metal. In the intermediate regions, both between Figure 28b and 28c and between 28d and 28e more complex spectral features are observed.[52]

A further point of interest is the mechanism of heat dissipation of a metal nanoparticle after energy has been deposited by irradiation with a short laser pulse at the surface plasmon frequency. In the same way as in bulk metal, the excited electrons first relax by electron–electron scattering to give a hot electron distribution. The hot electrons then equilibrate with the lattice by electron–phonon coupling. In the bulk metal the energy deposited by the laser then diffuses away from the excitation region by thermal conductivity, but in nanoparticles this route is not available: the energy has to be transferred from the hot electron–phonon system to the environment. A study of heat dissipation for colloidal gold particles in aqueous solution showed that energy relaxation is highly non-exponential. The relaxation times were found to be independent of the initial temperature of the particles, but they scale with the square of the diameter, *i.e.* the higher the surface area the slower is heat dissipation. For very small particles (~ 4 nm) the time scale for heat dissipation is comparable to the time scale of electron–phonon coupling, which amounts to several picoseconds. This implies that significant energy transfer to the environment can occur before electrons and phonons have reached equilibrium.[55] At this point the results represent phenomenological findings which have to await a fundamental understanding (Figure 29).

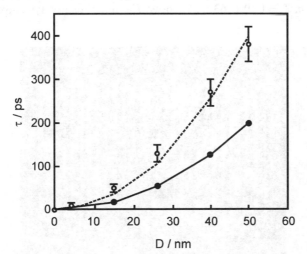

Figure 29 *Characteristic stretched exponential time constant for energy dissipation versus diameter of colloidal gold nanoparticles in aqueous solution. Open circles: experimental data, full dots: calculated dependence. The dashed line shows a fit to the data assuming a parabolic dependence of τ on the diameter* (Reproduced with permission from Ref. 55. Copyright (2002) American Chemical Society)

6   A Semiconductor Quantum Dot Electronic Device

Todays lithographic technologies permit the preparation of nanostructures with an amazing precision, as exemplified in Figure 30 by the well-defined array of silicon nanopillars with *ca.* 100 nm diameter and 5 μm height, obtained by a dry etching procedure in high density plasmas.[56]

The nanopillars displayed in Figure 30 are demonstrations of impressive technology, but these simple structures are of limited use and interest. Nanoelectrical devices have much more complex structures that consist of layers of different composition, and to be of interest they need to be contacted individually. Here we will explain briefly some of the basics of such an electrical device consisting of a 2-D few-electron quantum dot located in the semiconductor heterostructure pillar which is shown in the micrograph in Figure 31a. A full description is found in a review by Kouwenhoven *et al.*[57] The quantum dot consists of an undoped $In_{0.05}Ga_{0.95}As$ disk of 12 nm thickness along its growth direction z and *ca.* 500 nm diameter. It is sandwiched between two non-conducting AlGaAs barrier layers, separating it from the conducting layers above and below which serve as the source and drain contacts (Figure 31b).

The dimensions and compositions of the heterostructure were chosen such that the bottom of the conduction band in the quantum well is 32 meV below the Fermi level E_F of the contacts (Figure 31c). Taking into account zero-point energy the $n_z = 0$ (and $n, \ell = 0,0$) ground state is still below E_F, so that it contains electrons without application of a gate voltage, while the excited state $n_z = 1$ (and $n, \ell = 0,0$) is 63 meV above E_F and thus always empty. Since the

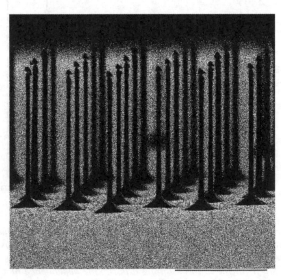

Figure 30   *Silicon nanopillars with a high aspect ratio*
(With permission by E. van der Drift from www.nanopicoftheday.org/ 2004Pics/June2004/nanopillars.htm)

thickness of the dot is much smaller than its diameter the system can in good approximation be regarded as a 2-D system, so that there are many n, ℓ states between $n_z = 0$ and $n_z = 1$. These are not shown in Figure 31c but on an expanded energy scale in Figure 31d as a function of a magnetic field along the z-direction of the dot. For a diameter of ~ 500 nm these states can hold about 80 electrons at zero gate voltage. A negative voltage applied to a metal gate around the pillar squeezes the diameter of the dot's lateral dimension (dotted lines in Figure 31b). With increasing voltage the energy levels are shifted up

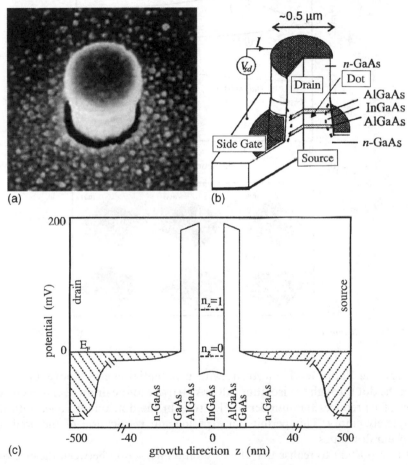

(a)

(b)

(c)

Figure 31 *Semiconductor heterostructure pillar containing a quantum dot. (a) Electron micrograph of the pillar which has a width of ca. 500 nm. (b) Schematic diagram of the electrical device, a three-terminal field-effect transistor. (c) Composition of the graded heterostructure and corresponding course energy diagram. (d) Calculated detailed single-particle energy diagram relative to the $n_z = 0$ ground state versus magnetic field. (e) Examples of the square of some single-particle wavefunctions for different quantum numbers (n, ℓ)*
(Reproduced with permission from Ref. 57)

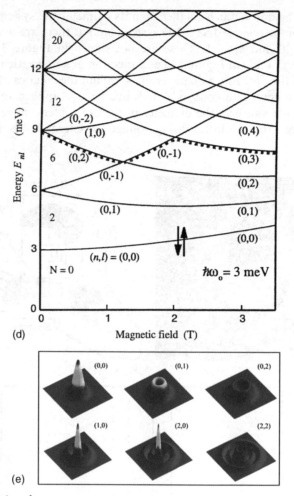

(d)

(e)

Figure 31   *continued*

through the Fermi level. Electrons with a potential energy above E_F tunnel from the dot through the isolating AlGaAs barrier to the drain so that a current flows. This reduces the number of electrons in the dot, one by one, until it is completely empty. This permits an experimental verification of the predicted quantum dot states.

It is important to realise that the Coulomb interaction between the electrons in the well is strongly screened by polarisation of the free electron distribution in the source and the drain, since the latter are only ~ 10 nm away, much less than the dot diameter. The energies of the electronic states of this 2-D pseudo-atom are therefore given to a good approximation as single-particle energies in a close to 2-D harmonic potential (Figure 31d). The squared wavefunctions of some of these states are displayed in Figure 31e. They remind very much of wavefunctions of real atoms.

Key Points

- The number of states in an electronic band of a delocalised system is proportional to the number of atoms in the system. For partially filled bands the Kubo gap (*i.e.* the HOMO–LUMO or SOMO–LUMO gap) is therefore a strong function of the size of the system.
- When the thermal energy falls below the Kubo gap this has important consequences for conductivity and magnetism of a system.
- For very small systems the band width increases with the number of member atoms. This may lead to overlap of neighbouring bands with increasing system size. It is essential for elements with $(s)^2$ valence electron configuration (*e.g.* for Mg, Zn, Hg) where the s and p bands overlap only for systems exceeding a minimum size. Small systems are noble-gas-like and therefore weakly bound and insulating, while larger systems are metallic.
- The relevant question about a system is not 'Is it a metal?' but rather 'When is it a metal?'
- The band gap of small semiconductor particles reflects quantum confinement and scales therefore approximately as R^{-2}.
- Ionisation potential and electron affinity reflect the availability of electrons to form chemical bonds or to participate in redox reactions. By choosing the proper cluster size these properties can be tuned by several eV between the values for the bulk material and the individual atom.
- For a bulk metal the value of the ionisation potential equals that of the electron affinity. It is called *work function*.
- By proper choice of materials and particle size semiconductor nanoparticles and heterostructures can be designed, which provide optimum band energies for desired charge separation or electron transfer processes. The photophysical and photochemical processes of these nanoparticles are analogous to those which are well known for dye molecules.
- Semiconductor quantum dots or pseudo-atoms can be prepared in large numbers by chemical self-assembly processes or engineered on a more individual basis by lithographic techniques. In the latter case they can also be wired up to make quantum electronic devices.
- Metallic particles give rise in the optical part of the spectrum not only to inter-band gap excitations but also to *collective excitations* of conduction electrons. They represent standing waves and are called *plasmon resonances*.

General Reading

- J. Jortner, Cluster size effects, *Z. Physik D*, 1992, **24**, 247.
- P.P. Edwards, R.L. Johnston and C.N.R. Rao, On the size-induced metal-insulator transition in clusters and small particles, in *Metal Clusters in Chemistry*, vol. 3, P. Braunstein, L.A. Oro and P.R. Raithby (eds), Wiley, Weinheim, 1999.

- P.P. Edwards and F. Hensel, Metallic oxygen, *ChemPhysChem*, 2002, **3**, 53.
- B. von Issendorff and O. Chesnovsky, Metal to insulator transitions in clusters, *Annu. Rev. Phys. Chem.*, 2005, **56**, 549.
- The metal-non-metal transition in macroscopic and microscopic systems, P.P. Edwards, R.L. Johnston, C.N.R. Rao, D.P. Tunstal and F. Hensel (eds.) *Philos. Trans. R. Soc. London, Ser. A*, 1998, **356**, Number 1735 (special issue), 3–278.
- M.J. Kelly and R.J. Nicholas, The physics of quantum well structures, *Rep. Prog. Phys.* 1985, **48**, 1699.
- W.P. Halperin, Quantum size effects in metal particles, *Rev. Mod. Phys.*, 1986, **58**, 533.
- A.D. Yoffe, Low-dimensional systems: quantum size effects and electronic properties of semiconductor microcrystallites (zero-dimensional systems) and some quasi-two-dimensional systems, *Adv. Phys.*, 1993, **42**, 173.
- U. Kreibig and M. Vollmer, Optical Porperties of Metal Clusters, Springer Series in Materials Science, Vol. 25, Springer, Berlin, 1995.

References

1. R.L. Whetten, D.M. Cox, D.J. Trevor and A. Kaldor, *Surf. Sci.*, 1985, **156**, 8.
2. A.D. Yoffe, *Adv. Phys.*, 1993, **42**, 173.
3. A.D. Alivisatos, *Science*, 1996, **271**, 933.
4. B. Winkler, C.J. Pickard, V. Milman, W.E. Klee and G. Thimm, *Chem. Phys. Lett.*, 1999, **312**, 536.
5. R.C. Haddon, *Acc. Chem. Res.*, 1992, **25**, 127.
6. A.F. Hebard, *Physics Today*, November 1992, 26.
7. W. Andreoni, F. Gygi and M. Parrinello, *Chem. Phys. Lett.*, 1992, **189**, 241.
8. J. Hrusák and H. Schwarz, *Chem. Phys. Lett.*, 1993, **205**, 187.
9. E. Roduner and I.D. Reid, *Chem. Phys. Lett.*, 1994, **223**, 149.
10. M.J. Kelly and R.J. Nicholas, *Rep. Prog. Phys.*, 1985, **48**, 1699.
11. H.C. Manoharan, C.P. Luz and D.M. Eigler, *Nature*, 2000, **403**, 512.
12. P.P. Edwards, R.L. Johnston and C.N.R. Rao, *On the size-induced metal-insulator transition in clusters and small particles*, in *Metal Clusters in Chemistry*, vol. 3, P. Braunstein, L.A. Oro and P.R. Raithby (eds), Wiley, Weinheim, 1999.
13. P.P. Edwards and F. Hensel, *ChemPhysChem*, 2002, **3**, 53.
14. R. Kubo, *J. Phys. Soc. Jpn.*, 1962, **17**, 975.
15. (*a*) A.L. Kuzemsky, Fundamental Principles of the Physics of Magnetism and the Problem of Localized and Itinerant Electron States, preprint JINR, Dubna, 2000, E17-2000-32, p. 22; (*b*) A.L. Kuzemsky, *Int. J. Modern Phys.*, 2002, **16**, 803.
16. J. Jellinek and P.H. Acioli, *J. Phys. Chem. A*, 2002, **106**, 10919.

17. R. Busani, M. Folkers and O. Cheshnovsky, *Phys. Rev. Lett.*, 1998, **81**, 3836.
18. M.B. Knickelbein, *J. Chem. Phys.*, 1993, **99**, 2377.
19. T. Schmauke, R.-A. Eichel, A. Schweiger and E. Roduner, *Chem. Phys. Phys. Chem.*, 2003, **5**, 3076.
20. A. Kawabata, *J. Phys. Soc. Jpn.*, 1970, **29**, 902.
21. A. Chatelain, J.-L. Millet and R. Monet, *J. Appl. Phys.*, 1976, **47**, 3670.
22. J. Michalik, D. Brown, J.-S. Yu, M. Danilczuk, J.Y. Kim and L. Kevan, *Phys. Chem. Chem. Phys.*, 2001, **3**, 1705.
23. J. Korringa, *Physica*, 1950, **16**, 601.
24. J.J. van der Klink, in *Physics and Chemistry of Finite Size Systems: From Clusters to Crystals*, P. Jena, B.K. Rao and S.N. Khanna (eds), Kluwer Academic, Dordrecht, 1992, 537.
25. H.E. Rhodes, P.-K. Wang, C.D. Makowka, S.L. Rudaz, H.T. Stokes, C.P. Slichter and J.H. Sinfelt, *Phys. Rev. B*, 1982, **26**, 3569.
26. M. Bastea, A.C. Mitchell and W.J. Nellis, *Phys. Rev. Lett.*, 2001, **86**, 3108.
27. W.J. Nellis, R. Chau, P.P. Edwards and R. Winter, *Z. Phys. Chem.*, 2003, **217**, 795.
28. U. Simon, in *Nanoparticles. From Theory to Application*, G. Schmid (ed), Wiley-VCH, Weinheim, 2004.
29. J. Park, A.N. Pasupathy, J.I. Goldsmith, C. Chang, Y. Yaish, J.R. Petta, M. Rinkoski, J.P. Sethna, H.D. Abruna, P.L. McEuen and D.C. Ralph, *Nature*, 2002, **417**, 722.
30. W. Liang, M.P. Shores, M. Bockrath, J.R. Long and H. Park, *Nature*, 2002, **417**, 725.
31. S. De Francesci and L. Kouwenhoven, *Nature*, 2002, **417**, 701.
32. S.K. Nielsen, M. Brandbyge, K. Hansen, K. Stokbro, J.M. van Ruitenbeek and F. Besenbacher, *Phys. Rev. Lett.*, 2002, **89**, 066804.
33. M.A. Continentino, B. Elschner and G. Jakob, *Europhys. Lett.*, 1995, **31**, 485.
34. R.H. Whetten, D.M. Cox, D.J. Trevor and A. Kaldor, *Phys. Rev. Lett.*, 1985, **54**, 1494.
35. B. von Issendorff and O. Chesnovsky, *Annu. Rev. Phys. Chem.*, 2005, **56**, 549.
36. A.W. Dweydari and C.B.H. Mee, *Phys. Status Solidi A*, 1973, **17**, 247.
37. K. Rademann, *Ber. Bunsenges. Phys. Chem.*, 1989, **93**, 653.
38. K.J. Taylor, C.L. Pettiette-Hall, O. Chesnovsky and R.E. Smalley, *J. Chem. Phys.*, 1992, **96**, 3319.
39. K.R.S. Chandrakumar, T.K. Ghanty and S.K. Ghosh, *J. Phys. Chem. A*, 2004, **108**, 6661.
40. J. Lüning, J. Rockenberger, S. Eisebitt, J.-E. Rubensson, A. Karl, A. Kornowski, H. Weller and W. Eberhardt, *Solid State Commun.*, 1999, **112**, 5.
41. A.I. Ekimov and A.A. Onushchenko, *JETP Lett.*, 1981, **34**, 345.
42. A.I. Ekimov and Al.L. Efros, *Acta Phys. Polonica.*, 1991, **79**, 5.
43. Al. L. Efros and A.L. Efros, *Soviet. Phys. Semicond.*, 1982, **16**, 772.

44. N. Chestnoy, T.D. Harris, R. Hull and L.E. Brus, *J. Phys. Chem.*, 1986, **90**, 3393.

45. P.V. Kamat, *Electron transfer processes in nanostructured semiconductor thin films*, in *Nanoparticles and Nanostructured Films*, J.H. Fendler (ed), Wiley-VCH, Weinheim, 1998.

46. A. Eychmüller, *J. Phys. Chem. B*, 2000, **104**, 6514.

47. A. Mews, A. Eychmüller, M. Giersing, D. Schooss and H. Weller, *J. Phys. Chem.*, 1994, **98**, 934.

48. D.V. Talpin, I. Mekis, S. Götzinger, A. Kornowski, O. Benson and H. Weller, *J. Phys. Chem. B*, 2004, **108**, 18826.

49. P.A. Sant and P.V. Kamat, *Phys. Chem. Chem. Phys.*, 2002, **4**, 198.

50. C.F. Landes, M. Braun and M.A. El-Sayed, *J. Phys. Chem. B*, 2001, **105**, 10554.

51. C.A. Leatherdale, W.-K. Woo, F.V. Mikulec and M.G. Bawendi, *J. Phys. Chem. B*, 2002, **106**, 7619.

52. U. Kreibig and M. Vollmer, *Optical Properties of Metal Clusters*, Springer Series in Materials Science, Vol. 25, Springer, Berlin, 1995.

53. G. Mie, *Ann. Phys.*, 1908, **25**, 377.

54. F. Mafuné, J. Kohno, Y. Takeda and T. Kondow, *J. Phys. Chem. B*, 2002, **106**, 7575.

55. M. Hu and G.V. Hartland, *J. Phys. Chem. B.*, 2002, **106**, 7029.

56. M.A. Blauw, T. Zijlstra, R.A. Bakker and E. van der Drift, *J. Vacuum Sci.*, 2000, **18**, 3453.

57. L.P. Kouwenhoven, D.G. Austing and S. Tarucha, *Rep. Prog. Phys.*, 2001, **64**, 701.

CHAPTER 5
Magnetic Properties

1 A Brief Primer on Magnetism

Magnetism has played an important role in solid-state physics as well as in chemistry over decades. The challenge of increasing the information density in data storage media and the development of new technological processes which permit the production of smaller and smaller magnetic particles has kept the subject alive. In the course of these studies, it became apparent that the dependence of magnetic properties on size and dimensionality is among the most spectacular size effects, so the subject cannot be omitted here. Since most chemists are not so familiar with magnetic phenomena, it seems appropriate to give a brief introduction into the phenomenology of those forms of magnetism which are most relevant to nanoparticles. We will see that it is again the reduced coordination of the constituent atoms near the surface which is primarily responsible for these effects.

1.1 The Basic Parameters

The relevant properties are the strength of the applied magnetic field, H, the magnetic flux density inside a medium, B and the magnetisation, M. They are related *via*

$$B = \mu_0 \left(H + M \right) = \mu_0 \mu_{\mathrm{r}} H \tag{1}$$

where $\mu_0 = 12.566 \times 10^{-7}$ Vs A^{-1} m^{-1} is the vacuum permeability and μ_{r} is the dimensionless relative permeability which gives the enhancement factor of B over $\mu_0 H$ due to magnetisation of the medium. In the regime where the magnetisation scales linearly with H, it is useful to define the magnetic susceptibility, X

$$M = XH = \left(\mu_{\mathrm{r}} - 1 \right) H \tag{2}$$

Magnetisation and susceptibility are given per volume, per mass unit or per mol. μ_{eff} is the effective atomic or molecular magnetic moment.

Two principal measurements are normally carried out: $M(H)$, the magnetisation as a function of applied field at a given temperature, and $M(T)$, the magnetisation as a function of temperature at a fixed field. When the $M(H)$ curve is linear and reversible, it is often X that is used for interpretation.

1.2 Curie Paramagnetism

A system of non-interacting magnetic moments due to unpaired electron spins combined with their orbital angular moments shows paramagnetic behaviour. M scales as a function of H according to the Brillouin function $B_j(a)$ which saturates at high fields and low temperatures,

$$M \propto B_j(a) = \left(\frac{2j+1}{2j}\right) \coth\left(j \cdot a + \frac{a}{2}\right) - \frac{1}{2j}\coth\left(\frac{a}{2}\right) \qquad (3)$$

with $a = g\mu_B H/k_B T$, where g is the Landé factor, and j is the quantum number for the total (spin plus orbital) angular momentum. For paramagnetic ions, the ionic magnetic moment is sufficiently small so that for not too high fields and not too low temperature, M increases to a good approximation linearly with H. The behaviour is illustrated for four temperatures in Figure 1 for Mn^{2+}, which has $j = 5/2$, diluted in CdS. The agreement of the data with a 5/2 Brillouin function and the absence of a hysteresis demonstrate that the spin 5/2 manganese ions form an ensemble of non-interacting spins which are doped into a non-magnetic host material.

The molecular susceptibility of paramagnetic samples scales with temperature as

$$X_m(T) = \frac{C_m}{T} \qquad (4)$$

Normally, $X_m^{-1}(T)$ is plotted, giving rise to a straight line through the origin and with slope C_m^{-1}. The molar Curie constant C_m is related to the effective

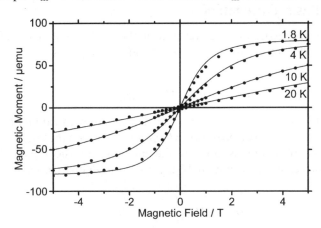

Figure 1 *Magnetic moment of a sample of Mn doped cadmium sulfide nanoparticles, showing the 5/2 Brillouin type sigmoid saturation curves that characterises a paramagnetic state. At 10 and 20 K, the dependence is already linear up to a magnetic field of 5 T*
(Adapted from Ref. 1. Reproduced with permission from the PCCP Owner Societies)

atomic magnetic moment μ_{eff} as

$$\mu_{\text{eff}} = \left(\frac{3k_B C_m}{N_L \mu_0}\right)^{1/2} \tag{5}$$

Thus, if the molar concentration of spins is known μ_{eff} can be determined. For $S = 1/2$ in the absence of an orbital angular momentum μ_{eff} adopts a value of 1.73 μ_B.

1.3 Curie–Weiss Paramagnetism

An exchange interaction between the different magnetic moments attempts to align adjacent moments in the same or in opposite direction. The effective field that is seen by any magnetic moment is then given by the sum of the externally applied and an internal (molecular) field. Below a critical temperature this results in a long-range order, but above it one finds Curie–Weiss paramagnetism given by

$$X_m(T) = \frac{C_m}{T - \theta}, \qquad (T > T_C, T_N) \tag{6}$$

$T_C = \theta$ is called the Curie–Weiss temperature. It is related to the strength of the interaction between moments. Its sign depends on whether the interaction prefers parallel alignment, which below T_C results in ferromagnetism ($\theta > 0$), or antiparallel alignment, which leads to antiferromagnetism ($\theta < 0$). A plot of X_m^{-1} versus T leads to straight lines which are offset by θ (see Figure 2), while the slope is related to μ_{eff} as given for the paramagnetic case (Equation 5).

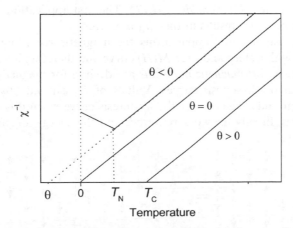

Figure 2 *Schematic drawing of the inverse susceptibility for the cases of Curie paramagnetism ($\theta = 0$), and Curie–Weiss paramagnetism. For $\theta < 0$, the behaviour turns antiferromagnetic below T_N, for $\theta > 0$ X is history-dependent and has no meaning in the ferromagnetic phase below T_C*

For $\theta > 0$, X diverges at $T = \theta$. For $\theta < 0$, there is no divergence at positive temperature, but the low temperature long-range order breaks down near the Néel temperature, $T_N \approx |\theta|$

1.4   Antiferromagnetism

In antiferromagnetic materials, the magnetic moments of nearest neighbours are oriented antiparallel to each other. Very often, these systems can be considered as consisting of two interpenetrating sublattices, one of which has its magnetic moment pointing up and the other one down (Figure 3). The net magnetic moments which result from antiferromagnetic coupling below T_N are relatively small so that the $M(H)$ behaviour resembles that of a paramagnet, and X remains a meaningful property. The details of antiferromagnetic couplings can be quite complex, and some of the complications which are particularly important for nanomaterials will be introduced in Chapter 5.2. The antiferromagnetic phase is a long-range ordered state.

1.5   Ferromagnetism and Ferrimagnetism

In ferromagnets, all magnetic moments are long-range ordered parallel to each other within a magnetic domain. This results in quite large domain moments which can be fully aligned in a sufficiently large external field, and in very distinct $M(H)$ and $M(T)$ curves. Both are non-reversible, so one has to follow certain procedures to obtain history-independent intrinsic properties. Figure 4 shows the typical symmetric hysteresis behaviour of $M(H)$ that is obtained on cycling the external field to values beyond the magnetic fields where the magnetisation reaches its *saturation* value, $\pm M_s(T)$. The curves cross zero external field at the *remanent magnetisations*, $\pm M_r(T)$, and the magnetisations become zero at the *coercive fields*, $\pm H_c(T)$. The first application of an external field starts at zero and results in the *virgin curve*.

Permanent magnets and applications for magnetic recording require *hard ferromagnets* with a steep and wide $M(H)$ curve and therefore high values of M_r and H_c, whereas *soft magnetic materials* are desired for magnetic transformer cores with narrow hysteresis curves. Values of H_c can vary by about seven orders of magnitude. The area of the hysteresis curve when it is drawn as flux density $B(H)$ is directly the work required, and the energy lost as heat, over a

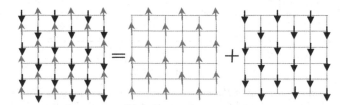

Figure 3   *An antiferromagnetic lattice can be decomposed into two interpenetrating sublattices*
(Reprinted from Ref. 2 with permission from Oxford University Press)

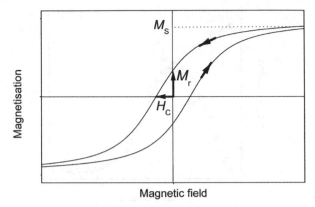

Figure 4 *Hysteresis curve of the magnetisation M(H) of a ferromagnetic material. The intrinsic properties are the saturation magnetisation M_s, the remanent magnetisation M_r and the coercive field H_c. The arrows give the cycling direction*

complete hysteresis cycle. The ratio M_r/M_s in principle yields information about the symmetry of the magnetocrystalline anisotropy.

When the number of moments that contribute to magnetisation is known, the effective moment per atom, μ_{eff}, can be evaluated from the maximum value of the saturation magnetisation M_s that is measured at low temperature where all the moments are aligned parallel. There is also a *spontaneous magnetisation* of the domains below T_C that occurs in the absence of an external field, but the domains are randomly oriented so that there is no macroscopic magnetisation. It reaches its maximum value at $T = 0$, and it decreases at first slowly and then goes to zero rapidly with increasing temperature, giving rise to critical behaviour near T_C.

A second way of accounting for the history dependence of a magnetic material is to do a zero-field cooled/field cooled (ZFC/FC) set of measurements. This is done by annealing the sample at a temperature which makes sure that any macroscopic magnetisation has vanished, and then cooling it to the lowest measurement temperature at $H = 0$. Then, a field is applied, and the magnetic moment is measured as a function of temperature up to the highest desired temperature. This is the ZFC part. For the FC part, the sample is cooled in the desired field to the lowest temperature and again measured as a function of temperature. At least in the case of superconductors, it is recommended that the values are collected on cooling.

A typical set of curves is shown in Figure 5. One important piece of information that is derived from it is the temperature, T_s, where the two curves split, called the splitting or irreversibility temperature. Since all measurements are conducted in the same magnetic field, the two curves are identical above T_s, but below this point the completely disordered state is trapped in the ZFC measurement so that it starts off near zero where only those terms contribute to magnetisation which are not related to the aligned magnetic domains. Up to T_s, this branch is measured in a non-equilibrium state, so that one should be aware

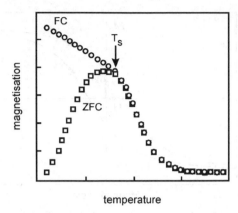

temperature

Figure 5   *A schematic view of a set of magnetisation curves obtained on cooling in the magnetic field in which the measurements are performed (FC, upper curve) and after cooling in zero field (ZFC, lower curve) and measuring in the same field as for the FC case. This is a convenient way of determining over which temperature range the sample shows irreversible behaviour. T_s is the splitting temperature*

of the time scale of a possible relaxation into equilibrium. The measurements may therefore depend to some extent on the time needed for the experiment, in particular near T_s.

Ferrimagnetic materials show behaviour in between those of ferromagnets and antiferromagnets. The two sublattices of an antiferromagnet may be inequivalent, so that their magnetisations do not cancel out. The two components of the net magnetisation may have quite different temperature dependences. The overall temperature dependence may therefore be quite complicated and does not follow Curie–Weiss behaviour. The phenomenon is observed often for mixed oxides of iron.

1.6   Molecular Magnets

Single-molecule magnets belong to the more general class of compounds called exchange coupled clusters in which a small number of paramagnetic transition metal ions are linked together by simple bridges such as O^{2-}, HO^-, H_3CO^-, $RCOO^-$ or also halogens. These bridges serve as a pathway for a magnetic superexchange that leads to ferromagnetic or antiferromagnetic exchange coupling which is often in the order of $1–100$ cm^{-1}. This cluster core is shielded by organic ligands so that there is no significant coupling between neighbouring clusters.

Very high magnetic moments are often found in molecular transition metal clusters such as the iron wheel (Fe$_8$), or the Mn$_{12}$-acetate [Mn$_{12}$O$_{12}$-(CH$_3$COO)$_{16}$·4H$_2$O] which has an $S = 10$ ground state (Figure 6a). The Mn$_{12}$-acetate relaxes so slowly at low temperatures that individual molecules are reminiscent of bulk magnets so that such systems are often called single-molecule magnets. If a molecule is magnetised by application of a field, this

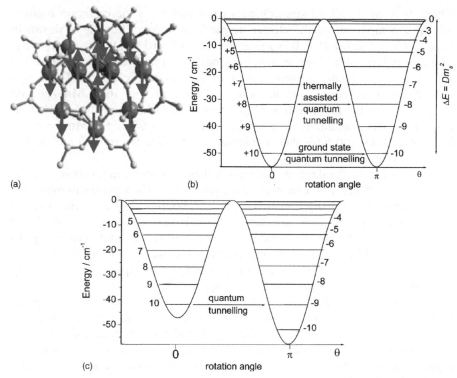

Figure 6 *(a) Mn$_{12}$-acetate molecular magnet. The green balls in the core are Mn^{4+} ions (S = 3/2, occupying four out of the eight corners of a cube), the orange balls at the periphery represent Mn^{3+} ions (S = 2), and the small balls are O^{2-} ions and acetate ligands. Within the core and within the periphery the coupling is ferromagnetic, but between core and periphery it is antiferromagnetic, giving rise to a S = 10 ground state. (b) Energy level diagram for the 21 eigenstates $-10 \leq m_s \leq +10$. The sin$^2\theta$ curve represents the classical energy curve for rotation of the total spin in a uniaxial crystal field. (c) Energy level diagram in the presence of an external magnetic field of 0.5 T parallel to the crystal field axis*

magnetisation keeps for approximately 2 months at 2 K, or for 40 years (!) at 1.5 K, even after switching off the magnetic field.[3] The slow relaxation, despite the fact that it is of purely molecular origin, gives rise to hysteresis effects, similar to that of bulk magnets. It therefore becomes in principle possible to store information on a single molecule, which is of high technical interest. The prerequisite for such behaviour is a high magnetic anisotropy due to crystal field.

In the simplest case in which we have axial symmetry with negligible transverse anisotropy and an external magnetic field B applied parallel to the symmetry axis of the system (also called the *easy axis*) the energies are given by

$$\varepsilon(M_s) = Dm_s^2 + g\mu_B B m_s \tag{7}$$

where D is the zero-field splitting parameter and m_s is the quantum number for the z-component of the cluster magnetic moment. The case is shown in Figures 6b and 6c for $D < 0$ in a magnetic field of zero and 0.5 T, respectively. The energy level of a given m_s value belongs to a quantised direction θ of the spin angular momentum with respect to the crystal field axis. There is a potential barrier between the states $m_S = 10$ and $m_S = -10$ of Mn_{12}-acetate which prevents the free rotation of magnetisation at low temperature. Classically, the barrier height is $|D|S^2$, which amounts to 64 K for Mn_{12}-acetate. Systems with $D > 0$ have their minimum energy state for $m_s = 0$ and do not exhibit a double well.

A reason for the fundamental interest in these molecular magnets is given by the quantum effects which show up in the relaxation. The system can overcome the barrier *via* a thermally activated process or, because the states with $\pm m_s$ are degenerate in the absence of an external magnetic field, by *quantum tunnelling of magnetisation*.[3,4] At low temperatures where only the ground state is significantly populated this occurs directly between the $m_S=10$ and $m_S=-10$ states, but at somewhat elevated temperatures it may also happen as a thermally assisted process between excited states (Figure 7).[4]

The quantised magnetisation curves with their vertical sections are characteristic for molecular magnets. They represent an ensemble average over many molecules. At a given sweep rate, the largest width is obtained for the lowest temperature where thermally assisted tunnelling contributes less so that a higher field is needed to reduce the barrier height above the populated state and to accelerate tunnelling.

By comparison, a single domain classical magnet would show a single vertical section and thus a near-rectangular hysteresis loop, while multi-domain classical magnets exhibit more continuous curves. Under certain conditions, the above hysteresis curves can be short circuited by vertical lines. The corresponding process which is termed *avalange* relaxes the non-equilibrium situation at once. This occurs when the heat released on passage of a number of magnets to their lower energy well cannot be dissipated, so that a sufficient amount of thermal energy becomes available that further magnets can cross the barrier.

The organic ligands in Mn_{12}-acetate serve to couple the magnetic ions within the cluster, but at the same time they act as spacers, which isolate one cluster from the others so that they may be treated to a good approximation as individual, non-interacting clusters.

To date the species with the highest magnetic moment ($S = 51/2$) is a cyano-bridged $Mn_9^{II}Mo_6^{V}$-molecular cluster.[5]

1.7   Superparamagnetism

At low temperature, all spins of single domain clusters are coupled to quite large magnetic moments which have a fixed direction in a cluster. For changing this direction one has to overcome the anisotropy energy. These clusters therefore display ferromagnetic or ferrimagnetic behaviour which is recognised by the presence of hysteresis in a magnetisation curve.

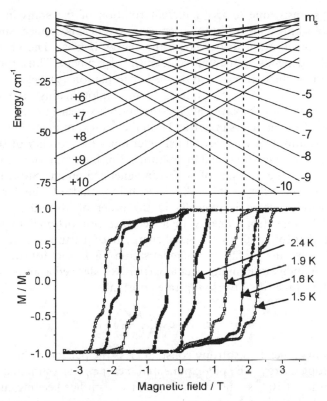

Figure 7 *Energy level diagram of the Mn$_{12}$-acetate molecular magnet as a function of magnetic field (upper) and magnetisation hysteresis loops given in units of the saturation magnetisation m$_s$ as a function of applied magnetic field that is swept at a fixed rate. The curves are determined from SQUID measurements at different temperatures*

In a higher temperature regime where the thermal energy exceeds the anisotropy energy (this critical temperature is called the *blocking temperature*, T_b), these coupled moments behave like individual spins in paramagnetic materials. However, because the magnetic moments are so large these super-spins can much more easily align, they therefore saturate in a lower external field than the small uncoupled spins of paramagnetic materials. In contrast to the magnetic moments of molecular magnets these superspins behave classically, so that their angle with respect to the external field is not quantised but can take any value. Their saturation behaviour is then given by the Langevin function in place of the Brillouin function (Equation 3).

$$M = N\mu\left[\coth\left(\frac{3g\mu_B H}{k_B T}\right) - \left(\frac{k_B T}{g\mu_B H}\right)\right] \qquad (8)$$

Superparamagnetism is the most important form of magnetism in context with nanomaterials. It is characterised by a large reversible magnetisation without hysteresis, due to free coherent rotation of the spins in each grain above the blocking temperature. T_b decreases as a field is applied, since the field helps the magnetisation to overcome the energy barrier. The observation of blocking also depends on the time scale of the experiment. Thus, for relaxation due to thermal fluctuations as described in Equation (9) below, the inverse blocking temperature decreases linearly in a plot against the log of the frequency of dynamic experiments (ac-susceptibility measurements).[6]

One of the key quantities in particular in the context of magnetic recording is the magnetic anisotropy energy which determines the stability of the magnetisation direction against thermal fluctuations. For bulk materials, this energy is proportional to the volume of a ferromagnetic domain. Sufficiently small ferromagnetic particles with a diameter below that of the ferromagnetic exchange length, which is typically in the order of 10–100 nm, are single domains. The anisotropy energy then becomes proportional to the particle volume, so that for small particles the energy barrier can be small enough for thermal fluctuations to flip the magnetisation back and fourth.[7] Using the classical probability of jumping over a barrier Néel derived a simple relation for the relaxation time (the inverse flip rate constant):

$$\tau = \tau_0 \exp\left(\frac{KV}{k_B T}\right) \tag{9}$$

where K is the crucial anisotropy constant, V is the particle volume and KV is the anisotropy energy. The pre-exponential factor τ_0 was estimated by Néel to be in the order of 10^{-9} s, only in more recent years it has become customary to take it as 10^{-10} s. Since Néel, more advanced theories have been introduced,[7] but because of its simplicity, Equation (9) is still used very often for a first approximation.

An example illustrates the dramatic importance of the size effect: based on Equation (9) and on known anisotropy energies of iron an estimate gives a relaxation time for a spherical iron particle at 300 K of 0.1 s for a diameter of 23 nm, but of 10^9 s at 30 nm.[7] This makes clear that any dependence of the anisotropy constant on particle size is essential.

1.8   Other Forms of Magnetism

All materials have a *diamagnetic* component, which has its origin in atomic ring currents leading to an induced magnetisation antiparallel to the external field. It leads to a small negative susceptibility that is temperature independent. In closed shell molecules, this is the only contribution to magnetism. Since the diamagnetic moment scales with the square of the atomic radius, small metallic clusters that can be regarded as pseudo-atoms should be expected to exhibit a giant diamagnetism.

Conduction electrons in metals give rise to *Pauli paramagnetism* with a positive susceptibility of about three times the magnitude of the diamagnetic contribution. In most cases, it is constant to a good approximation down to very low temperature.

In systems with a non-magnetic ground state, a paramagnetic contribution can be generated by admixture of excited states with non-zero orbital magnetic moment. This orbital susceptibility is positive and temperature independent, and according to its discoverer is called *Van Vleck paramagnetism*.

Spin glasses represent a configuration of spins which are frozen into a more or less random pattern. This state is established below a distinct freezing or glass temperature, T_g, from a randomly fluctuating state. The most common signature of a spin glass is a difference between a FC and a ZFC experimental cycle below T_g.

Spin liquids differ from spin glasses in that no distinct T_g is observed, that is the spins, or a large fraction of them, remain dynamic down to the lowest temperatures.

2 The Concept of Frustration

On a simple basis, we might expect that clusters containing an even number of electrons should be closed shell systems, and thus diamagnetic, and odd electron clusters should have a single unpaired electron each, giving rise to paramagnetic behaviour. However, it is also common knowledge that systems with a sufficiently high symmetry have a degenerate ground state which may lead to a violation of this simple rule. For example, the oxygen molecule, a 'dinuclear even-electron cluster' has an electronic triplet and thus paramagnetic ground state. Furthermore, based on Hund's rule, transition metal atoms and ions with 2–8 valence electrons have a spin $S > 1/2$. For 4 to 7 d-electrons they occur with a *high spin* ground state, unless there is a lowering of symmetry due to crystal field effects that is accompanied by a sufficiently large energy splitting so that the ground state adopts a *low spin* configuration.

Small clusters of high symmetry give rise to similar phenomena. The geometry of an equilateral triangle and that of a square involve symmetry elements which lead to doubly degenerate states, those of tetrahedral symmetry to triple degeneracy. A chemist will tackle the situation of a four atom cluster of tetrahedral symmetry by drawing molecular orbitals, and if there are four valence electrons to fill into a triply degenerate molecular orbital one ends up with two unpaired electrons. If this symmetry is broken, it permits all electrons to be paired.

Extended solids are often built up from a large number of high symmetry fragments such as corner-sharing or edge-sharing equilateral triangles or tetrahedra. When there is no coupling between the unpaired electrons at the corner atoms this is a trivial case of Curie paramagnetism. The case of ferromagnetically coupled electron spins is still relatively straightforward. Antiferromagnetic coupling, however, can lead to a much more complex behaviour for which solid-state physicists have developed a new language involving the *concept of frustration*.

A cluster that is built from only a relatively small number of such high symmetry fragments represents an intermediate situation between that of a small molecule and an extended solid. An accurate description will be based on high level quantum chemical molecular orbital calculations. However, there will often be a large number of *frontier orbitals* (the highest occupied and lowest unoccupied MOs) and their energetic order will depend on how they are occupied with electrons, on the spin state, and on small changes in geometry. Such calculations may be extremely difficult. It may, therefore, be easier to approach the system based on the view of an extended solid, and then think about corrections which may be necessary due to the broken symmetry near the surface. It is again the surface with its low coordinated atoms and concomitant dangling bonds and unpaired electrons which is often responsible for unexpected magnetic behaviour of nanoparticles. To approach such systems, it is useful to first consider some of the basics of spin frustration.

Heisenberg's phenomenological Hamiltonian which describes the interaction between two spins i and j can be written as a scalar product of the spin operators

$$H_{ij}^{ex} = -2J_{ij}S_iS_j = -2J_{ij}(S_i^x S_j^x + S_i^y S_j^y + S_i^z S_j^z) \qquad (10)$$

where the exchange coupling J_{ij} is assumed to be isotropic (more generally, it should be represented by a tensor). It reflects that the energy is minimised for collinear (parallel for $J_{ij} > 0$, or antiparallel for $J_{ij} < 0$) spin alignments. For pure through-space point-dipolar interactions J_{ij} decreases with the third power of distance. The generally more important mechanisms of the through-bond exchange and superexchange (through an intermediate non-magnetic bridge atom) interaction are proportional to overlap integrals which decay exponentially with distance. In an ensemble of N spins, there are $N(N-1)/2$ pair-wise interactions. Because of the distance dependence of the exchange interaction, it is normally sufficient to consider only nearest neighbours (J_{nn}) and next-nearest neighbours (J_{nnn}), even though the through-bond interaction may complicate the picture considerably.

Cases where all short-range J_{ij} are positive can easily be satisfied and lead to a *ferromagnetic* spin order with all spins parallel. In contrast, *antiferromagnetic* short-range interaction leads to a much more complex situation with a rich palette of magnetic behaviour. It is subject to extensive current investigations on infinite periodic lattices.

The *spin dimensionality D* is given by the number of components which must be retained in the expanded form of Equation (10). The simplest form where only the z component is important is known as the *Ising model* ($D = 1$), while $D = 2$ stands for the $X-Y$ *model*, and $D = 3$ for the isotropic *Heisenberg model*. This spin dimensionality comes separately from the dimensionality of the lattice, d, and both are important for the realisation of long-range magnetic order. The *lattice dimensionality* depends on whether the magnetic structure elements are arranged in chains (1-d), planes (2-d) or in isotropic 3-d geometry.

Magnetic frustration represents a situation wherein a large fraction of magnetic sites in a lattice is subject of competing or contradictory constraints. When this situation arises purely from the topology of the lattice, it is termed *geometric frustration.*[8] The conditions of magnetic frustration are satisfied in many real materials, and the effects of frustration are present in the magnetic properties, but they are not always appreciated.

The classical example is based on an equilateral triangle which may be a corner- or edge-shared structure element of a magnetic lattice. The situation for Ising spins is depicted in Figure 8b. When the interaction of two neighbouring spins of the triangle is antiferromagnetic so that their spins are antiparallel in the ground state this condition can be satisfied only by two of the three spins. The third spin can meet this condition only with one but not with the other one of the two neighbours, *it is said to be frustrated.* This situation of energetic degeneracy of two orientations of the third spin which therefore has no preferential orientation is indicated with a question mark in Figure 8.

Four spins on a square lattice with $|J_{nn}| >> |J_{nnn}|$ (neglect of diagonal interactions) can all be satisfied, as shown in Figure 8a. When the spins are positioned on the corners of a tetrahedron (consisting of four edge-shared equilateral triangles) there is no possibility to satisfy more than two spins (Figure 8c). A more difficult situation arises when the exchange coupling across the diagonal of a square is of similar magnitude as along the edge, $J_{nn} \approx J_{nnn}$. For this possibly hypothetical case, we have two high energy (parallel spin) and four low energy interactions no matter how we arrange the spin orientations around the square.

The non-Ising cases where more than one term of Equation (10) has to be retained are also termed *compromise* or *non-collinear* spin structures. The 2-*D* case of an equilateral triangle and the 3-*D* case of a tetrahedron are shown in Figure 9. Here, the vector sum is zero for each structure element, and thus the antiferromagnetic constraint is satisfied without any uncoupled spins. This leads to a magnetic behaviour that is different from that of the configuration in Figure 8. Most frustrated systems which undergo true phase transitions to a long-range ordered state adopt such structures.[8]

The energy scale on which interesting phenomena occur is set by the exchange energy, $-2J_{ij}S^2 \approx kT$. A simple experimental measure of the exchange energy is provided by the Curie–Weiss temperature θ (Equation 6). In the absence of frustration, one expects the onset of strong deviations from the Curie–Weiss law at a critical temperature $T_c \leq |\theta|$, and the establishment of a long-range ordered state near T_c. For ferromagnetic order, this is observed to be valid to a good approximation, that is $\theta/T_c \approx 1$. For antiferromagnetic order things are more

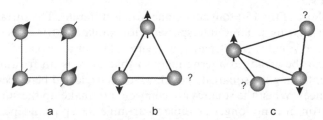

a   b   c

Figure 8 *Antiferromagnetic interaction of Ising spins on a square (a), an equilateral triangle (b) and a tetrahedron (c). The question marks indicate frustrated spins*

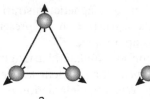 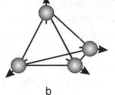

a   b

Figure 9 *Compromise non-linear magnetic structures for a D = 2 triangular arrangement (a) and a D = 3 tetrahedral arrangement (b)*

complex, and typical values of $|\theta|/T_c$ are in the range *between* 2 and 5.[8] Somewhat arbitrarily, $|\theta|/T_c > 10$ is often adopted as a criterion indicating the presence of frustration. This means that despite large negative values of θ frustrated systems show good Curie–Weiss behaviour down to low temperatures. Below T_c, susceptibility measurements of frustrated spin systems display a clear difference between the FC and a ZFC experimental cycle.

The Kagome lattice – the expression derives from the analogy to a distinctive Japanese basket wave pattern – is built up from planes of corner-shared triangles around hexagonal holes. Two possible compromise structures are shown in Figure 10. In all cases where experimental data exist, the structure 5.10a is found to describe the spin configuration, even though 5.10b was predicted to be favoured from both classical and quantum chemical theories.[8] It was anticipated that interplanar coupling plays a role in controlling the spin configuration.

The surface of a 13-atom icosahedron consists of 20 equilateral triangles. This leads to various frustrating interactions. An example out of many possible arrangements of the spins is given in Figure 11 (left panel). It is based on the Ising model and assumes that all the spins on the horizontal pentagon through B are oriented *down*, the ones on the plane through D are *up*. In this way, spin A and spin E can be accommodated so that they have no frustrating interaction, while the spin on the central atom C (not drawn) has the choice between equivalent orientations. It is the extra spin which gives rise to a paramagnetic contribution. For this arrangement, each spin on the planes B and D have frustrating interactions with two neighbours (the ones in the same plane) while three interactions are favourable. In total, this gives 20 frustrating interactions over the entire surface. The same number is found for other antiferromagnetic arrangements.

The topology of the 13-atom cuboctahedron is different. The surface consists of six corner-sharing squares, interspersed with equilateral triangles sharing the edges with the squares (right panel in Figure 11). On an isolated square, the four spins can be arranged alternatingly so that there is no frustrating interaction (except along the square diagonal, which is neglected here because of the larger distance). When the squares are connected to make up the surface of the cuboctahedron, it is no longer possible to arrange the spins independently. In the example of Figure 11, the spins on the equator through B are arranged alternatingly *up* and *down*, so there is no frustration on this hexagon. Three of

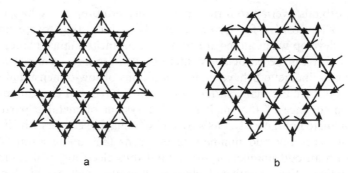

Figure 10 *Two conceivable compromise spin structures on the extended Kagome lattice (D = d = 2) (Reproduced from Ref. 8 with permission from the Royal Society of Chemistry)*

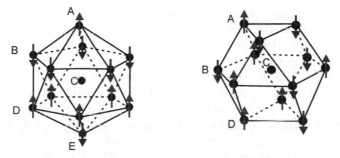

Figure 11 *Examples for possible antiferromagnetic arrangements of Ising spins on the surfaces of a 13-atom icosahedron and a 13-atom cuboctahedron, illustrating the concept of frustration. The spin on the central atom C is not drawn since it has no preferential orientation and can be regarded as an extra spin with paramagnetic behaviour*

the spins on the top and bottom triangles have to be *up* and three *down*, but it is impossible to arrange them in a way in which they do not introduce frustration, whereas the spin on atom C has the same distance to all 12 surface spins (six *up* and six *down*), so it has no preferential orientation and can be regarded as the extra spin with paramagnetic behaviour, as in the case of the icosahedron.

3   Magnetic Properties of Small Clusters

3.1   Theoretical Predictions

Ferromagnetism is normally found for the transition metal elements iron, cobalt and nickel, and for a few rare earth metals and their alloys. It is only for these elements that the atomic spin and orbital angular moment do not get lost when the orbitals overlap to form a bond. Furthermore, the exchange interaction has to be positive for ferromagnets. It is interesting to see how these properties develop when atoms are combined to linear chains, planes and 3-D clusters, and it will be particularly interesting to see how elements which are normally non-magnetic behave under conditions of low-dimensional arrangements where there are many surface atoms with a reduced coordination.

One consequence of the small size of a system of metallic particles was already predicted in the early 1960s by Kubo[9]: Imagine clusters which are built from atoms carrying odd numbers of electrons (*e.g.* alkali atoms). Neutral clusters with an odd number of atoms (and thus electrons) will exhibit Curie paramagnetism. Even numbered clusters, however, will exhibit the normal Pauli paramagnetism of metals, except at low temperature where additional spin pairing as a consequence of the small energy gap leads to susceptibilities below the Pauli value. For odd–even mixed ensembles this can lead to a complicated magnetic behaviour at low temperature. Moreover, the heat capacity can be reduced below the normal value at low temperatures when there is an energy gap above the ground state.

A Kubo gap in the order of kT will also strongly affect the electron relaxation which arises from thermal excitations involving reorientation of spins. For small particles, this should be observable at low temperature. Kubo estimates that the average energy spacing of electronic levels of an N-atom cluster is of the order of E_F/N (E_F is the Fermi energy).[9] For a cluster of 10^5 atoms this will amount to about 10^{-4} eV, which corresponds to the electron Zeeman level splitting in a field of 1 T, or an energy of kT at 1 K. For smaller clusters the splitting will be correspondingly larger.

Atomic magnetic moments were calculated based on an *ab initio* tight binding approximation for a large number of Fe, Co, Ni and V atoms arranged to linear chains, monoatomic planar layers, and bulk volume, in the latter case also as a function of cluster size.[10] The internuclear distances were fixed at the bulk equilibrium value. The results are displayed in Figures 12 and 13.

The results reveal that the single factor that dominates magnetism is the immediate environment of the magnetic atom under investigation, characterised by its coordination number. Specifically, we recognise the following principles:[10]

- There is a clear effect of lattice *dimensionality*. The magnetic moment decreases as the dimensionality increases from the atom to the linear chain, the planar layer and the bulk configuration. In the case of V, the bulk phase is non-magnetic, however, in a linear chain the moments are ferromagnetically coupled. As the dimensionality of the system decreases the probe atom has fewer nearest neighbours, that is the coordination number decreases. As a result the bands become narrower, giving rise to a higher magnetic moment.

- Second, there can be an effect of *symmetry*. Clusters of high symmetry are more likely to have a (multiply) degenerate ground state. Such states give rise to high spin configurations because of Hund's rule coupling, it is thus

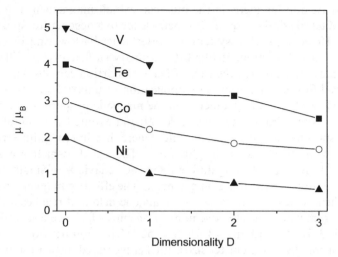

Figure 12 *Magnetic moments per atom in μ_B of V, Fe, Co and Ni for individual atoms (0-D), linear chains (1-D), planar layers (2-D) and bulk arrangements (3-D) at T = 0*
(Prepared from data in Ref. 10)

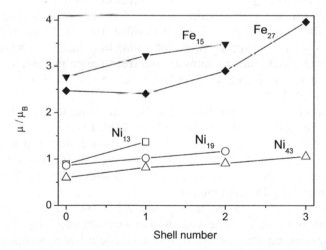

Figure 13 *Spin moments per atom in μ_B at T = 0 of atoms in the first, second and third shell around a central atom 0*
(Prepared from data in Ref. 10)

expected that a high symmetry cluster would carry a larger magnetic moment per atom than a corresponding cluster with reduced symmetry. Experimentally, this was confirmed for Fe_{13} clusters which occur with icosahedral as well as cuboctahedral structures. The number of unpaired spins in the icosahedral cluster was larger than in the cuboctahedral cluster with the same radial bond distance.[11] In principle, also clusters of elements which are non-magnetic in the bulk can exhibit magnetism. For example, Na_4 of tetrahedral or square geometry leads to a degenerate, spin 1 ground state. However, such a system will often react by lowering its energy *via* Jahn–Teller distortion. If this lifts geometry sufficiently, it will lead to a low spin case (spin 0 in the case of Na_4). Distortion can also be effected by external forces, for example by application of uniaxial stress.

- The effect of size and geometry on the magnetic properties of icosahedral and cuboctahedral clusters was calculated in Monte Carlo simulations for Ising spins located at the atomic sites where they interact antiferromagnet-ically with their nearest neighbours.[12] The results which are shown in Figure 14 reveal a striking difference in the behaviour of small clusters (13 and 561 atoms) depending on geometry. The effective magnetic moment of icosahedral clusters increases with magnetic field and reaches a saturation value from which the effective number of spins which are free to respond to field is 3 for $N = 13$ and 13 for $N = 55$.[12] This number does not increase proportionally as the cluster size is further increased, rather it remains at 13 up to $N = 147$ atoms, and then μ_{eff} drops to 0.07 μ_B per atom, one-third of its value for the smaller clusters. This reflects the cluster topology which at the surface of $N = 13$ icosahedral clusters consists of equilateral triangles so that two spins can couple antiferromagnetically while the third one cannot satisfy this requirement, it is frustrated (compare Figure 11).

- As opposed to the icosahedral clusters, the ones of cuboctahedral sym-metry show an entirely different behaviour. For the 13-atom cluster, μ_{eff} remains at a constant high value, indicating that the cluster has a non-zero spin ground state but the spins are not free to align parallel to the field as this is tuned up. Larger clusters behave essentially the same way as large icosahedral clusters.

- The existence of free and blocked spins is also visible in the behaviour of μ_B as a function of temperature (Figure 14b). Icosahedral clusters follow essentially a paramagnetic pattern, while μ_{eff} is constant at low tempera-ture for the cuboctahedral case. As T increases further the weakly bound spins in the cuboctahedral cluster become disordered, which is particularly obvious for the 13-atom cluster.

- Figure 13 reveals an interesting size effect with a *radial gradient* of magnetic moments. The central atom of a cluster which has the maximum coordination number has the lowest magnetic moment which approaches its bulk value when the cluster size goes up. The moment *increases* as the surface layer is approached where the coordination number is lowest, and the moment of the surface atoms is nearly independent of the cluster size.

Remarkably, spins of materials which are paramagnetic in the bulk under all conditions such as Rh, Pd and Ru were predicted to be coupled ferromagnet-ically in a cluster.[13] The results for small osmium and rhodium clusters on a silver (001) surface are given in Figure 15. It is seen that the average magnetic moment per atom depends quite sensitively on cluster size (for comparison: the magnetic moment of a single unpaired electron amounts to 1.73 μ_B). In the case

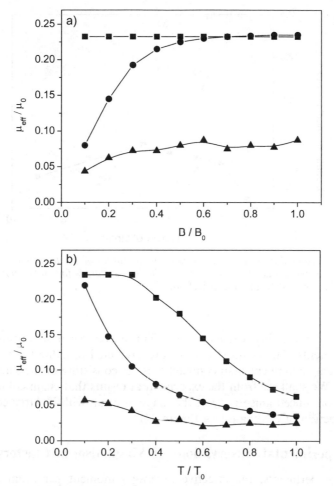

Figure 14 *Effective magnetic moment per atom in units of $\mu_0 = J/\mu_B$ (J is the exchange coupling) for cuboctahedral clusters of 13 atoms (squares) and icosahedral clusters of 13 (circles) and 561 (triangles) atoms as a function of field and a temperature of 0.2 ($T_0 = J/k_B$), and as a function of temperature in a magnetic field of 0.2 (in units of $B_0 = J/\mu_B$)*
(Prepared from data in Ref. 12)

of individual adatoms, the highest moments are naturally found for elements near a d^5 configuration, but for small islands this maximum shifts to d^6 or even d^7 elements.[14]

Free dimers of 4d atoms show large magnetic moments. Upon deposition onto a silver surface this moment decreases, and embedded in bulk silver it becomes negligible. This support or matrix effect was ascribed to strong hybridisation of the 4d wave functions with the Ag valence electrons, causing broadening of the density of states.[14] The moment per adatom for different configurations of 4d and 5d elements on an Ag(001) substrate surface usually

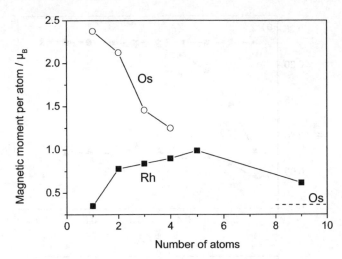

Figure 15   *Predicted average magnetic moment per atom of small osmium and rhodium clusters on an Ag(001) surface. The broken line is for an Os monolayer (Prepared from data in Ref. 14)*

decreases when the cluster size increases. This decrease is more pronounced for the 5d elements. The chain structures are predicted to exhibit a significantly larger magnetic moment than compact clusters consisting of the same number of atoms. We shall see from the experimental results that chain configurations lead to pronounced anisotropy of the magnetisation with its large component in the direction of the chain (see Figure 24).

3.2   Experimental Observations of Magnetism in Clusters

In beam experiments, the effective (average) moment per atom along the direction of the field, H, of a cluster of N atoms with total magnetic moment $N\mu$ in a field H is described classically by the Langevin function.[10]

$$\mu_{\text{eff}} = \mu\left[\coth\left(\frac{N\mu H}{3k_{\text{B}}T}\right) - \frac{k_{\text{B}}T}{N\mu H}\right] \tag{11}$$

For small fields, μ_{eff} depends linearly on H and N and is given by

$$\mu_{\text{eff}} = \frac{N\mu^2 H}{3k_{\text{B}}T} \tag{12}$$

The magnetic moment and its temperature dependence were measured for size-selected clusters of iron, cobalt and nickel in a molecular beam.[15] Figure 16 shows how the average magnetic moment develops for iron clusters. An individual neutral atom has two paired 4s electrons and four unpaired d electrons, thus, neglecting any contribution from the orbital angular momentum one would expect a magnetic moment of 4 μ_{B}. The plot in Figure 16 starts

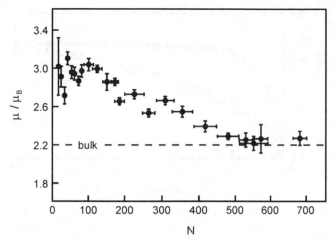

Figure 16 *Average (effective) magnetic moment per atom in units of μ_B for iron clusters at 120 K as a function of cluster size up to nearly 700 atoms* (Redrawn from Ref. 15. Copyright (1994) AAAS)

off at about 3 μ_B, for small clusters of *ca.* 30 atoms. This corresponds partly to the situation in the bulk where only one electron per atom is in the 4s band, but the bulk band structure is obviously not fully developed.[15]

For all three elements which are ferromagnetic in the bulk the average magnetic moment per atom is large for small clusters and decreases gradually towards the bulk value with increasing cluster size. Superimposed on the decay, there is an oscillating pattern with a period that corresponds roughly to one atomic layer for the cases of Co and Ni. For Fe, a crystal phase transition may complicate the situation. For Ni, the second layer from the surface was reported to be magnetically inactive; for Fe, the fourth layer appears to be antiferromagnetically coupled.[15]

Ferromagnetic ordering has been found experimentally for Rh_N ($N = 12$–32),[16] and Rh_{13} was predicted to have 21 unpaired electrons and thus a magnetic moment of 21 μ_B.[17] Fe, Cr and V, on the other hand, have the same atomic electronic structure and therefore already in small clusters also a similar magnetic behaviour as in the bulk.

It should be noted that the saturation magnetisation is reduced only slightly (by a few percent) for nanocrystalline Ni or Co in comparison with corresponding bulk polycrystalline references, but the shape of the hysteresis loop changes.[18] With decreasing grain size steeper and narrower loops are observed, and thus the saturation magnetisation can be reached at smaller external field (Figure 17). This agrees with the expectation that spins can reorient more easily in small clusters.

The blocking temperature of small particles of magnetite (Fe_3O_4) was studied and confirmed to depend on particle size by Bandow and Kimura[19] (Figure 18). T_b is expected to saturate when the particle size has reached the magnetic exchange length.

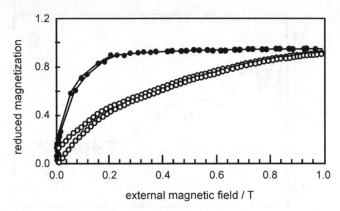

Figure 17 *Magnetisation relative to its saturation value for nanocrystalline Co of 20 nm (dots) and 73 nm (open circles) grain diameters at room temperature. For the smaller particles, the hysteresis has almost vanished*
(Redrawn with permission from Ref. 18. Copyright (2001) American Physical Society)

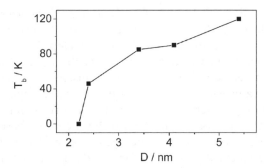

Figure 18 *Blocking temperature of ultrafine magnetite (Fe_3O_4) particles as a function of diameter*
(Based on data from Ref. 19)

An extremely strong superparamagnetic behaviour down to below 4 K was found for nanoparticles of gold and palladium, elements which are not normally magnetic.[20] Particles of both metals have a narrow size distribution with a mean diameter of 2.5 nm. The saturation magnetisation as estimated based on the data shown in Figure 19 amounts to 20 μ_B for Pd and to 30 μ_B per particle for Au, which demonstrates that there is a large number of uncompensated spins, probably near the surface (see Figure 13). It was concluded that at least the Au particles must be metallic. An anomalous magnetisation that increases with decreasing size and reaches a value of 6.3 μ_B per particle at a diameter of 2.6 nm diameter was also observed for monodisperse Pt nanoparticles.[21]

The above findings are in line with spin density functional calculations for neutral and anionic Pd_N clusters. For neutral clusters with $2 \leq N \leq 7$ they reveal

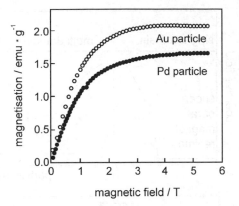

Figure 19 *Magnetisation of gold and palladium nanoparticles with a mean diameter of 2.5 nm at 1.8 K*
(Reprinted from Ref. 20 with permission from Elsevier)

a spin triplet ground state, and for $N = 13$ a spin nonet is predicted. The calculations confirm that most of the spin density is located at the cluster surface.[22]

For ferromagnetic particles isolated by a non-magnetic matrix, the coercive field becomes a maximum at the size where the particles become single domains. This is at a size in the order of the ferromagnetic exchange length, which is typically in the range of 10–100 nm. Nanocomposite materials with this particle size have large-scale commercial applications as hard magnets. Similar to conventional ferromagnets, the magnetisation direction in the absence of an external field is aligned along the easy axis determined by the matrix lattice and by the shape anisotropy of each particle. When the particle size is reduced further the energy required to rotate the magnetisation vector out of the easy direction diminishes, and at sufficiently small size thermal energy will be sufficient to do it. The array of isolated particles as a whole then loses its ferromagnetic properties and becomes *superparamagnetic*. The temperature at which the width of the hysteresis curve vanishes and above which superparamagnetism is observed is called *blocking temperature*. The smallest grain sizes of about 1 nm then become excellent soft magnetic materials with a magnetic permeability of close to 10^5.

A brief account of the size dependence of the temperature and size dependence of the coercive field was given by Herzer.[23] He showed that for grains smaller than the exchange length L_{ex}, the coercivity H_c and the permeability μ scale with the grain diameter d as

$$H_c \propto d^6 \quad \text{and} \quad \mu \propto H_c^{-1} \propto d^{-6} \tag{13}$$

while for grains larger than L_{ex} the dependence is

$$H_c \propto d^{-1} \quad \text{and} \quad \mu \propto d \tag{14}$$

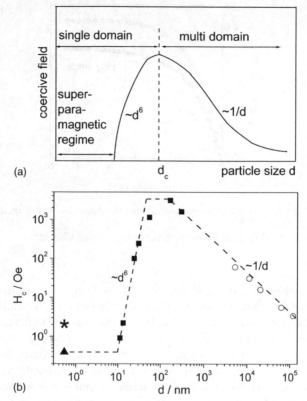

(a)

(b)

Figure 20   *Coercivity as a function of grain size for nanocrystalline magnetic matter. (a) Schematic behaviour. (b) Nanocrystalline alloys at room temperature: Fe–Co$_{0-1}$Nb$_3$(SiB)$_{22.5}$ (squares), 50% Ni–Fe (circles), amorphous Co (triangle), amorphous Fe (asterisk)*
(Data from Ref. 23. Copyright© 1990 IEEE)

The dependence was verified for various soft magnetic alloys,[23] and Figure 20 demonstrates that these magnetic parameters can vary over four orders of magnitude. The maximum coercivity is obtained for the maximum particle size d_c that is of single domain. Below a second critical value, the anisotropy energy is so small that the coercivity and thus the hysteresis disappear so that the particles adjust freely to the external field. This is the superparamagnetic regime.

While for larger particles the temperature dependence is moderate, it is quite dramatic for particles of a few nanometre diameter (Figure 21), and a crossover is observed at around 30 K.[24]

Magnetic iron oxide nanoparticles in the size range of 2–8 nm were investigated by López-Pérez *et al.*[25] using X-ray diffraction, TEM and magnetisation measurements. It was demonstrated that an evaluation of the magnetic saturation curves (Figure 22) based on the Langevin function for superparamagnetic particles with a log-normal size distribution gave a particle volume in agreement with that from the Scherrer formula and from TEM. The larger

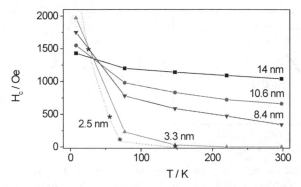

Figure 21 *Coercivity as a function of temperature for metallic iron particles of various core diameters inside a Fe oxide shell. The points are connected as a guide to the eye*
(Redrawn with permission from Ref. 24. Copyright (1992) by the American Physical Society)

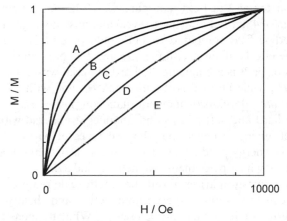

Figure 22 *Reduced magnetisation versus field at room temperature for iron oxide nanoparticles with a mean diameter of 7.3 nm (A), 5.6 nm (B), 3.6 nm (C), 2.4 nm (D) and 1.9 nm (E)*
(Reprinted with permission from Ref. 25. Copyright (1997) American Chemical Society)

particles appeared to saturate more easily, and the saturation value (M_s) decreased with decreasing particle size. Furthermore, FC and ZFC sets of curves were obtained. The temperature of irreversibility, given by the onset of splitting between the two curves, decreased from about 200 K for 7.3 nm particles to below 50 K at 1.9 nm (Figure 23). For superparamagnetic particles with a size distribution, this temperature equals the highest blocking temperature, T_b. It corresponds to the largest particle in the sample and reflects the known dependence on the particle volume (compare Figure 18).

An interesting observation arises from magnetic measurements of FePt bimetallic systems.[26–28] Even though bulk Pt is non-magnetic, it is able to enhance the

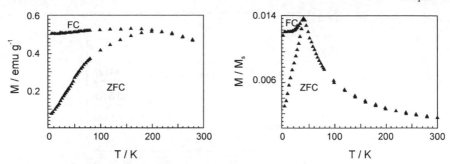

Figure 23 *Field cooling (FC) and zero-field cooling (ZFC) cycles of magnetisation measurements of iron oxide nanoparticles with a mean diameter of 7.3 nm (left) and 1.9 nm (right). The particles are superparamagnetic above the point where the two curves split*
(Selection reprinted with permission from Ref. 25. Copyright (1997) American Chemical Society)

saturation magnetisation in bulk FePt alloy by a factor of 1.56, and it may be by as much as a factor of 4 in FePt nanoparticles of 2.5 nm diameter.[26] At 10 K, Mössbauer data of 4 nm FePt particles reveal an unusually high and well-defined magnetic hyperfine field compared to FePt bulk or multilayer samples.[27] The interplay of the effects of surface, finite size and Fe core polarisation is not yet fully understood. FePt is a magnetically hard material with a large value of the coercive field H_c, while Fe_3Pt is soft with small H_c, and with a high magnetisation. Zheng *et al.*[28] developed magnetic nanocomposites *via* a self-assembled mixture of the hard and soft phases and obtained a material with an enhanced energy product, giving evidence that the two types of particles are exchange-coupled. A large energy product, related to the area of a hysteresis loop, is the figure of merit by which permanent magnets are judged.

An interesting question arises about the length scale required for ferromagnetic order. Clusters show superparamagnetic, and below the blocking temperature ferro- or ferrimagnetic behaviour. What happens when particles are divided finer and finer? It was shown that the blocking temperature drops to zero when the size of magnetite particles decreases to less than about 2 nm (Figure 18), and ferromagnetic resonance is only observed for particles larger than 3.4 nm.[19] Obviously, long-range spin order disappears. On the other hand, iron clusters display high spin character up to at least Fe_{17}. These findings do not seem to be compatible. It transpires that the concept of long-range spin-ordering is not adaptable to small particles; it may have to be replaced by a concept of fluctuating high spin states.[19]

4   Ferromagnetic Order in Thin Films and Monoatomic Chains

In bulk material, ferromagnetic domains have a size which is controlled by the ferromagnetic exchange length, and in the absence of alignment in an external field they are randomly oriented. In thin films with a diameter less than this exchange length, it is expected that the magnetic properties are severely affected, in particular, the orientation of the domains may be either in plane or perpendicular to the surface.

It is crucial that low-dimensional systems are prepared in a clean and reproducible way since surface contamination, oxidation, but also kinks and steps or defects arising from mechanical treatments may considerably influence the magnetic properties. Below, some examples illustrate that even for well-defined epitaxial thin films of iron the magnetisation behaviour is quite complex.

Bulk iron crystallises in two phases: up to 1184 K it is bcc, between 1184 and 1665 K it is fcc, and from there up to the melting point at 1809 K it is again bcc. Below its Curie temperature, $T_C = 1017$ K, it shows ferromagnetic order. Epitaxial iron crystallises in its bulk bcc structure on Ag(100), independent of the thickness of the epitaxial film. For films thicker than 5 monolayers (ML), T_C adopts its bulk value, but for thinner films T_C drops linearly to a value of less than 400 K for the monolayer.[29] At $T = 30$ K the 3 and 4 ML films are perpendicularly magnetised at remanence, whereas thinner and thicker films have their magnetisation in plane. On the surface of Cu(100), epitaxial iron is stabilised in fcc structure up to a thickness of 15 ML. It is found to have a ferromagnetic ground state, and the 3 and 5 ML films have a T_C of 390 K which is considered to be the transition temperature for bulk fcc iron, while the single ML film has $T_C = 230$ K.[30] Near T_C the saturation magnetisation is known to drop linearly with temperature. At room temperature the 3- and 5-ML films showed no measurable remanence, but at 30 K hysteresis loops with coercive fields of 350 and 650 Oe, respectively, were observed. The easy direction of magnetisation is perpendicular to the plane of the film. The 1-ML film showed no hysteresis even at low temperature.

Ultrathin epitaxial Fe films on Au(111) are magnetised in plane, but small amounts of Co in FeCo alloy films are sufficient to induce strong perpendicular magnetisation below three monolayers, with a pronounced striped domain structure.[31] In view of their potential applications in magnetic memories and in sensors it is essential to understand these materials.

Ultrathin iron films on a Cu(001) surface also exhibit stripe patterns with opposite perpendicular magnetisation in neighbouring domains. More interestingly, it was found that the low temperature stripe domain structure transforms into a more symmetric structure on heating, however, at even higher temperature and before the loss of magnetic order, a re-occurrence of the less symmetric stripe phase is found.[32] Such inverse transitions with lower symmetry at higher temperature are very rare.

A pioneering experiment involving arrays of sufficiently separated *one-dimensional chains* of cobalt atoms which were obtained by decoration of the regular step structure of a Pt(997) surface demonstrated the existence of both short- and long-range ferromagnetic order.[33] The experiment was interpreted in terms of thermally fluctuating segments of the chain of ferromagnetically coupled atoms. Below a threshold temperature, a long-range ferromagnetically ordered state evolved, owing to the presence of an anisotropy barrier. Characterisation of the cobalt chains revealed large localised orbital moments and correspondingly large

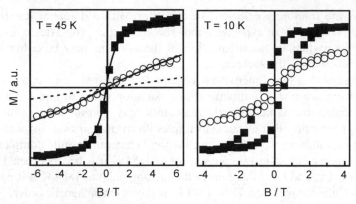

Figure 24   *Magnetisation curves of an array of monoatomic linear chains of cobalt atoms.*
            The magnetisation along the axis is clearly much larger along the axis than
            perpendicular to it, revealing its anisotropy. The absence of hysteresis at 45 K
            demonstrates its superparamagnetic character, while at 10 K the system is
            clearly ferromagnetic
            (Reproduced from Ref. 33 with permission from the journal)

magnetic anisotropy energies compared with two-dimensional films and with bulk
Co. As seen from the results of theoretical calculations presented in Figure 12, the
atomic magnetic moment is expected to increase continuously as the atomic
coordination is reduced in passing from the bulk *via* monoatomic plains to chains
and finally to the individual atom. The coordination is expected to influence in
particular the local orbital moment because of its sensitivity to the crystal field.
This was verified in the experiment, where the orbital moment was enhanced by
about a factor of five over that of bulk cobalt.

   Figure 24 shows magnetisation curves for the Co chains parallel and close to
perpendicular to the chain direction. At $T = 45$ K, the observed behaviour is
that of a 1-*D* superparamagnetic system, consisting of spin blocks of about 15
Co atoms within the chains which have an estimated uninterrupted length of 80
atoms in the average. The magnetic anisotropy energy amounts to 2.0 ± 0.2
meV per atom, which represents an increase by a factor of 50 over that of bulk
hcp Co and by a factor of 14 over that of a Co monolayer on Pt(997). When the
temperature is lowered below the blocking temperature of $T_b \approx 15$ K hysteresis
behaviour develops, giving evidence of a long-range ferromagnetically ordered
state with alignment of the magnetisation in the easy axis direction. It is a bit
uncertain how much the magnetisation is enhanced by polarisation of the non-
magnetic Pt support, as it is known to occur in PtFe alloys and clusters, but
there was no evidence of a coupling between adjacent chains that could be
transmitted through the support.

   Analogous experiments with double, triple and quadruple chains and also
monolayers revealed an interesting oscillatory behaviour of the easy axis of
magnetisation, the magnetic anisotropy energy and the coercive field as the system
evolves from 1- to 2-*D*-like.[34] In contrast to the trend expected for increasing
coordination of the Co atoms, the anisotropy energy and the coercive field

decrease sharply in going from the single to the double chain, then rise again in the triple wires before converging to monolayer values. The hard axis of magnetisation was always found to be along the wire direction, and the easy axis in a plane perpendicular to it, but at a variable angle.

It is interesting in this context that iron which is a typical ferromagnetic material turns antiferromagnetic in a monolayer on W(001).[35] This contrasts with the opposite case where non-magnetic materials such as Au, Pd and Pt become magnetic in nanoparticles (see Figure 19), and it illustrates again how dramatically different from bulk behaviour materials can be at low dimensions.

We have already seen in Chapter 5, Section 3.1 that there is an interaction of a magnetic transition metal guest atom with its support, an Ag(001) surface. Its magnetic moment is predicted to decrease compared with that of a free atom, and embedded in silver it disappears completely. In contrast, the magnetic moment of single cobalt atoms and nanoparticles on a Pt(111) surface was found to increase strongly due to the combination of the unquenched orbital moments and strong spin–orbit coupling induced by the platinum substrate.[36]

A very valuable and up to date method for the determination of magnetic properties of nanomaterials is *X-ray magnetic circular dichroism* (XMCD).[37] The method combines sensitivity to magnetic polarisation with element specificity and surface sensitivity. It uses left and right circularly polarised light which is swept through the absorption edges of the surface adatoms to be investigated. The XMCD signal is the difference between the two measurements along the magnetic field axis. It is strongly spin dependent and permits the separation of spin and orbital moments by application of the sum rules.[38]

5 Finite Size Effects in Magnetic Resonance Detection

5.1 Nuclear Magnetic Resonance

The properties that are measured in nuclear magnetic resonance of small metal particles are primarily the *Knight shift K* and the nuclear spin-lattice relaxation time T_1. While T_1 is related to electron spin fluctuations and thus to *dynamic susceptibility*, the Knight shift is proportional to the *static spin susceptibility*. In the same sense as the susceptibility, it can be represented as a sum of different contributions. For non-transition metals, the dominant term arises from the Fermi contact interaction at the nucleus. However, for transition metals there is a core polarisation term which is most important and opposite in sign compared with the Fermi contact term. For bulk Pt K amounts to -4.3%. In addition, there may be a small nuclear spin–electron orbital interaction.

As a consequence of the dependence on spin susceptibility NMR spectra of small metal particles may be expected to display temperature dependence attributable to quantum size effects. The susceptibility of even-electron particles normally goes to zero at low temperature whereas that of odd electron particles diverges. Since there will normally be a distribution of particle size this will lead to a distribution of resonance frequencies. Moreover, since the *Fermi contact*

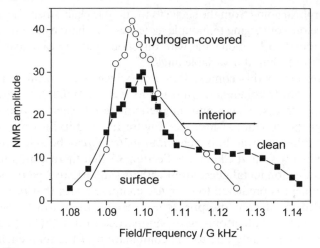

Figure 25   *^{195}Pt NMR spectra obtained point-by-point with an anatase supported Pt catalysts of 1.7 nm diameter and a Pt loading of 4.4 wt.%* (Based on data in Ref. 43)

interaction probes the spin susceptibility locally, NMR spectra will also reflect gradients, local inhomogeneity and Friedel oscillations near the surface, even at high temperature. These latter effects are expected to scale as the surface-to-volume ratio, that is with d^{-1}, but they will be temperature independent.

A very careful and extensive study was performed on Pt clusters with dispersions between 4 and 60% by the Slichter group. The ^{195}Pt NMR line shape depends sensitively on the state of the cluster surface, that is whether they are measured as synthesised, cleaned, exposed to air or to hydrogen.[39] The T_1 relaxation was measured as a function of the position across the broad, inhomogeneous line and showed a pronounced peak of weakest spin lattice relaxation for the resonance frequency that corresponds to Pt atoms at the cluster surface.[40] Evidence was presented also for a relation between T_1 and T_2 relaxation times. Furthermore, it was shown that there is a microscopic spatially oscillating variation of the Knight shift over the cluster radius.[41]

Small metal particles such as they occur in catalysis have Pauli-paramagnetic properties that are governed both by the density of states at the Fermi energy and by the local density of states, that is the intensity of the corresponding wave functions at each site. Van der Klink carried out NMR studies of small Pt and Ag metal particles.[42] The results for a 1.7 nm diameter Pt cluster are given in Figure 25 for the clean and the fully hydrogen covered surface.[43] The signal is extremely broad, covering about 6% of the NMR field, so that the spectra have to be recorded point by point. The large variation in the Knight shift between an atom at the surface and one in the bulk is ascribed to the variation in electronic structure. The surface peak was identified previously at 1.100 G kHz^{-1} at the low field edge of the spectrum, a position clearly very different from the 1.138 G kHz^{-1} bulk position.[44] For these clusters, the Korringa relation is obeyed, supporting the view that all Pt atoms are in a metallic environment. However, measurements on zeolite supported Pt particles of 0.9 nm diameter indicated that more than 50% of the particles are in a non-metallic environment.[42] It was suggested that the metal–non-metal transition occurs somewhere in the diameter range $0.7 < d < 1.2$ nm, depending to some extent also on adsorbates on the surface. Hydrogen adsorption changes significantly the spectrum, indicating that the density of states near the Fermi level is a sensitive function of the chemical state of the surface.

5.2 Electron Spin Resonance

ESR is probably the most direct and unambiguous approach to measure the electronic susceptibility of small particles and therefore particularly suitable for investigating the quantum size effect.[45] The parameters of interest are the resonance position or g value shift, the line width, the longitudinal relaxation time and the integrated intensity (the double integral of the derivative line) which scales with susceptibility. The ESR spectrum of atoms or ions is split by the *nuclear hyperfine interaction*, providing that it is sufficiently large, permitting the identification of the species. Also, small clusters with a well-defined structure give rise to observable hyperfine splitting (see below).

For light bulk metallic elements, *conduction electron ESR* is observed with a single-line signal intensity that is expected to be constant down to low temperatures on a scale given by the Kubo gap, $T \leq \delta/k_B$. Below this temperature the onset of Curie dependence should be seen for odd valence elements. The peak-to-peak width of a line with Lorentzian shape is given by

$$\Delta H_{pp} = \frac{2(\Delta g_\infty)^2}{\sqrt{3}\gamma_e \tau_R} \tag{15}$$

where Δg_∞ is the bulk g value shift from the free electron value, due to spin–orbit interaction, and $\tau_R = f(T)$ is the resistivity relaxation time.[45] It is associated with electron-phonon scattering and shows monotonic temperature dependence. For light elements, it scales with the inverse temperature above 77 K.

For particles smaller than the skin depth at the ESR experimental frequency, for typical metals in the order of 0.5–3 μm, there is an additional contribution to the resistivity relaxation rate, τ_R^{-1}, due to electron spin scattering from the particle surface, so that we have[45]:

$$\tau_R^{-1} = \frac{v_F}{d} + f(T) \tag{16}$$

It contributes to the ESR line width as

$$\Delta H_{pp} = \frac{2(\Delta g_\infty)^2}{\sqrt{3}\gamma_e} \left[f(T) + \frac{v_F}{d} \right] \tag{17}$$

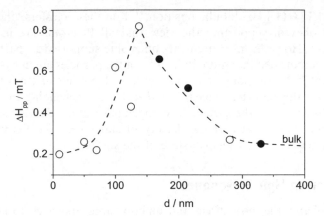

Figure 26   *ESR line width at 293 K as a function of size of Li particles deposited on a glass*
plate and covered with paraffin
(Reproduced with permission from Ref. 47)

For nanoparticles, this size-dependent term dominates over $f(T)$, and ΔH_{pp} becomes inversely proportional to the diameter. However, Equation (16) was derived for classical scattering. When the particle is sufficiently small the classical picture must be modified to take care of the discrete energy level spectrum of a system. Kawabata[46] found a corresponding solution which looks analogous to Equation (17) and differs only by a factor $h\nu_e/\delta$, where ν_e is the electron Zeeman frequency and δ is the Kubo gap:

$$\Delta H_{pp} = \frac{2(\Delta g_\infty)^2 h\nu_e}{\sqrt{3}\gamma_e\delta}\left[f(T) + \frac{\nu_F}{d}\right] \qquad (18)$$

Since δ varies as d^{-3} and $f(T)$ can be neglected at low temperature this means that the line width should now scale as the square of the particle size and thus *decrease* rapidly for small metallic particles. It seems that for small lithium particles the quantum limit was observed for particles below approximately 130 nm, whereas above it the line width showed classical behaviour (Figure 26). It is now commonly believed that this manifestation of the quantum size effect is the reason for facile observation of conduction electron ESR also for nanoparticles of heavier elements which would otherwise have very broad lines with a width in the order of a fraction of a Tesla.[45]

ESR spectra have been detected for ionic clusters of alkali metals in zeolites, such as Na_4^{3+} and K_4^{3+} as well as mixed alkali clusters in sodalite and in faujasites.[48] At low concentrations, the hyperfine splitting is resolved, proving the equivalence of the four alkali cores which share the unpaired electron. At higher concentrations, the electrons interact so that the hyperfine structure is washed out by fast spin exchange. When each sodalite cage which has a free inner diameter of 0.6 nm accommodates one of these clusters this gives rise to a

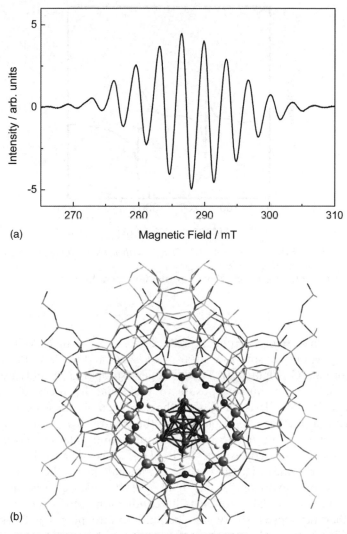

(a)

(b)

Figure 27   *X-band ESR spectrum of D_2-reduced Pt in NaY zeolite and suggested structure of $Pt_{13}D_{12}{}^+$ placed in the 12-ring window*
(From Ref. 51, Cover page of the issue; Reproduced by permission from the PCCP owner Societies)

body-centred cubic lattice of 2×10^{21} unpaired electrons per cm^3. $Na_4{}^+$ couples antiferromagnetically so that below a Néel temperature of 50 K a long-range ordered phase forms.[49] Other examples include well-defined clusters of silver and nickel in zeolites.[50]

More recently, a cluster consisting of 12 equivalent Pt atoms was observed in the zeolites KL, sodium faujasite (NaY) and sodalite.[51] It is covered entirely

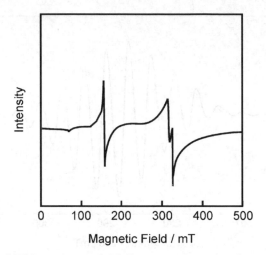

Figure 28 *X-band EPR spectrum of Pd clusters with a mean diameter of 2.8 nm in a polymer matrix at 4 K*
(Reprinted from Ref. 52 with permission from Elsevier)

with chemisorbed *H* or *D* from the reduction step, and in NaY it gives rise to a nearly isotropic spectrum near $g = 2.36$ and with a Pt hyperfine coupling of 68.8 G (Figure 27). The hydrogen isotopes are readily exchanged at room temperature, upon exposure to molecular hydrogen of the other isotopic composition. It is likely that the 12 Pt atoms form an icosahedral cluster with a further atom placed in the centre that is not spectrally resolved because of negligible hyperfine interaction. The composition was therefore suggested to correspond to $Pt_{13}H_{12}^{+}$. The well-resolved hyperfine structure demonstrates that the particle is molecule-like rather than metallic. The chemisorbed hydrogen atoms bind one Pt valence electron each so that they are not available for metallic conduction. Moreover, it is uncertain whether the cluster size would be sufficient to permit a transition to the metallic state in the absence of chemically bound surface hydrogen atoms.

X-band EPR spectra were obtained for Pd and Ni_xPd nanoclusters of a few nanometre diameters, stabilised in a polymer matrix.[52] Figure 28 shows a spectrum obtained with pure Pd. The main feature at $g = 2.01$ (*i.e.* lower than the *g*-value of bulk Pd of 2.2) is accompanied by an abnormally strong second harmonic or *half field line*, indicating that the cluster may actually be a triplet state. Even though the present cluster of 2.8 nm consists of many more atoms it reminds us of the prediction by Moseler *et al.*[53] that Pd clusters of up to seven atoms have a triplet ground state while a cuboctahedral 13-atom cluster was predicted to have a spin nonet ground state. Furthermore, the ESR spectrum shows quite a sharp line at $g = 2.00$ (compare also Ref. 54).

Key Points

- Conventional bulk magnetic properties are collective phenomena. In contrast, molecular magnets reflect single-molecule magnetism without coupling to neighbouring magnetic molecules.
- Molecular magnets are quantum systems which exhibit discrete steps in their magnetisation hysteresis curves.
- A high coordination number furnishes spin pairing. Uncompensated surface spins therefore become increasingly important in going from large clusters (3-D) to planar (2-D) and linear (1-D) systems.
- Atoms and molecules chemisorbed at the surface of a cluster pin unpaired surface electrons, thereby reducing the number of uncompensated spins and thus the magnetic moment of the cluster.
- Small cluster of non-magnetic materials such as Pd, Au and Pt adopt high magnetic moments, mostly due to uncompensated surface spins.
- A maximum coercive field is attained for the maximum size of a cluster that is of single magnetic domain. This occurs when the particle dimension is in the order of the magnetic exchange length.
- Single domain clusters behave like 'superspins' and show superparamagnetic behaviour as long as their interaction energy is below kT so that they respond independently to an external field and show no measurable hysteresis behaviour.
- It is the magnetic anisotropy energy which has to be overcome to reorient the magnetisation of a single domain or of a molecular magnet. This energy is proportional to the single domain volume and enters exponentially in the relaxation time.
- For antiferromagnetically coupled spin systems of high symmetry frustration plays an important role.

General Reading

- J.E. Gordan, Geometrically frustrated materials, *J. Mater. Chem.*, 2001, **11**, 37.
- P. Jena, S.N. Khanna and B.K. Rao (eds), *Physics and Chemistry of Finite Size Systems: From Clusters to Crystals*, vol. I, Kluwer Academic, Dordrecht, 1992.
- J.L. Dormann and D. Fiorani (eds), *Magnetic Properties of Fine Particles*, *North-Holland Delta Series*, Elsevier, Amsterdam, 1992.
- S. Blundell, in Magnetism in Condensed Matter, Oxford Master Series in Condensed Matter Physics, Oxford University Press, Oxford, 2001.
- W. Wernsdorfer, Classical and quantum magnetization studied in nanometer-sized particles and clusters, *Adv. Chem. Phys.*, 2001, **118**, 99.

- W.P. Halperin, Quantum size effects in metal particles, *Rev. Modern Phys.*, 1986, **58**, 533.
- S.N. Khanna and A.W. Castleman (eds), *Quantum Phenomena in Clusters and Nanostructures*, Springer, Berlin, 2003.

References

1. Ch. Barglik-Chory, Ch. Remenyi, C. Dem, M. Schmitt, W. Kiefer, Ch. Gould, Ch. Rüster, G. Schmidt, D.M. Hofmann, D. Pfisterer and G. Müller, *Phys. Chem. Chem. Phys.*, 2003, **5**, 1639–1643.
2. S. Blundell, *Magnetism in Condensed Matter, Oxford Master Series in Condensed Matter Physics*, Oxford University Press, Oxford, 2001.
3. D. Gatteschi, R. Sessoli and A. Cornia, *Chem. Commun.*, 2000, 725 (feature article).
4. L. Thomas, F. Lionti, R. Ballou, D. Gatteschi, R. Sessoli and B. Barbara, *Nature*, 1996, **383**, 145.
5. J. Larionova, M. Gross, M. Pilkington, H. Andres, H. Stöckli-Evans, H.U. Güdel and S. Decurtins, *Angew. Chem.*, 2000, **112**, 1667.
6. B.J. Jönsson, T. Turkki, V. Ström, M.S. El-Shall and K.V. Rao, *J. Appl. Phys.*, 1996, **79**, 5063.
7. A. Aharoni, in *Magnetic Properties of Fine Particles*, J.L. Dormann and D. Fiorani (eds), *North-Holland Delta Series*, Elsevier, Amsterdam, 1992.
8. J.E. Greedan, *J. Mater. Chem.*, 2001, **11**, 37.
9. R. Kubo, *J. Phys. Soc. Jpn.*, 1962, **17**, 975.
10. S.N. Khanna and P. Jena, in *Physics and Chemistry of Finite Size Systems: From Clusters to Crystals*, P. Jena, S.N. Khanna and B.K. Rao (eds), vol. I, Kluwer Academic, Dordrecht, 1992, 709.
11. B.I. Dunlap, *Phys. Rev. A*, 1989, **41**, 399.
12. B.V. Reddy and S.N. Khanna, in *Physics and Chemistry of Finite Size Systems: From Clusters to Crystals*, P. Jena, S.N. Khanna and B.K. Rao (eds), vol. I, Kluwer Academic, Dordrecht, 1992, 799.
13. M.S. El-Shall and A.S. Edelstein, in *Nanomaterials: Synthesis, Properties and Applications*, A.S. Edelstein and R.C. Cammarata (eds), Institute of Physics, Bristol, 1996.
14. P. Jena, S.N. Khanna and B.K. Rao, *Physics of clusters and cluster assemblies*, in *Theory of Atomic and Molecular Clusters*, J. Jellinek (ed), Springer, Berlin, 1999.
15. I.M.L. Billas, A. Châtelain and W.A. de Heer, *Science*, 1994, **265**, 1682.
16. A.J. Cox, J.G. Louderback and L.A. Bloomfield, *Phys. Rev. Lett.*, 1993, **71**, 923.
17. B.V. Reddy, S.N. Khanna and B.I. Dunlap, *Phys. Rev. Lett.*, 1993, **70**, 3323.
18. R. Przenioslo, R. Winter, H. Natter, M. Schmelzer, R. Hempelmann and W. Wagner, *Phys. Rev. B*, 2001, **63**, 54408.

19. S. Bandow and K. Kimura, *Z. Phys. D*, 1991, **19**, 271.
20. Y. Nakae, Y. Seino, T. Teranishi, M. Miyake, S. Yamada and H. Hori, *Physica B*, 2000, **284**, 1758.
21. Y. Yamamoto, T. Miura, Y. Nakae, T. Teranishi, M. Miyake and H. Hori, *Physica B*, 2003, **329–333**, 1183.
22. M. Moseler, H. Häkkinen, R.N. Barnett and U. Landman, *Phys. Rev. Lett*, 2001, **86**, 2545.
23. G. Herzer, *IEEE Trans. Magn.*, 1990, **26**, 1397.
24. S. Gangopadhyay, G.C. Hadjipanayis, B. Dale, C.M. Sorensen, K.J. Klabunde, V. Papaefthymiou and A. Kostikas, *Phys. Rev. B*, 1992, **45**, 9778.
25. J.A. López-Pérez, M.A. López Quintela, J. Mira, J. Rivas and S.W. Charles, *J. Phys. Chem. B*, 1997, **101**, 8045.
26. T. Schmauke, M. Menzel and E. Roduner, *J. Mol. Catal.: A Chem.*, 2003, **194**, 211.
27. B. Stahl, J. Ellrich, R. Theissmann, M. Ghafari, S. Bhattacharya, H. Hahn, N.S. Gaibhiye, D. Kamer, R.N. Viswanath, J. Weissmüller and H. Gleiter, *Phys. Rev. B*, 2003, **67**, 014422.
28. H. Zheng, J. Li, J.P. Liu, Z.L. Wang and S. Sun, *Nature*, 2002, **420**, 395.
29. M. Stampanoni, A. Vaterlaus, M. Aeschlimann and F. Meier, *Phys. Rev. Letts.*, 1987, **59**, 2483.
30. D. Pescia, M. Stampanoni, G.L. Bona, A. Vaterlaus, R.F. Willis and F. Meier, *Phys. Rev. Letts.*, 1987, **58**, 2126.
31. R. Zdyb and E. Bauer, *Phys. Rev. B*, 2003, **67**, 134420.
32. O. Portmann, A. Vaterlaus and D. Pescia, *Nature*, 2003, **422**, 701.
33. P. Gambardella, A. Dallmeyer, K. Maiti, M.C. Malagoli, W. Eberhardt, K. Kern and C. Carbone, *Nature*, 2002, **416**, 301.
34. P. Gambardella, A. Dallmeyer, K. Maiti, M.C. Malagoli, S. Rusponi, P. Ohresser, W. Eberhardt, C. Carbone and K. Kern, *Phys. Rev. Lett.*, 2004, **93**, 077203.
35. A. Kubetzka, P. Ferriani, M. Bode, S. Heinze, G. Bihlmayer, K. von Bergmann, O. Piezsch, S. Blügel and R. Wiesendanger, *Phys. Rev. Lett.*, 2005, **94**, 087204.
36. P. Gambardella, S. Rusponi, M. Veronese, S.S. Dhesi, C. Grazioli, A. Dallmeyer, I. Cabria, R. Zeller, P.H. Diederichs, K. Kern, C. Carbone and H. Brune, *Science*, 2003, **300**, 1130.
37. G. Schütz, W. Wagner, W. Wilhelm, P. Kienle, R. Zeller, R. Frahm and G. Materlik, *Phys. Rev. Lett.*, 1987, **58**, 737.
38. B.T. Thole, P. Carra, F. Sette and G. van der Laan, *Phys. Rev. Lett*, 1992, **68**, 1943.
39. H.E. Rhodes, P.-K. Wang, H.T. Stokes, C.P. Slichter and J.H. Sinfelt, *Phys. Rev. B*, 1982, **26**, 3559.
40. H.E. Rhodes, P.-K. Wang, C.D. Makowka, S.L. Rudaz, H.T. Stokes, C.P. Slichter and J.H. Sinfelt, *Phys. Rev. B*, 1982, **26**, 3569.

41. H.T. Stokes, H.E. Rhodes, P.-K. Wang, C.P. Slichter and J.H. Sinfelt, *Phys. Rev. B*, 1982, **26**, 3559.
42. J.J. van der Klink, Small metal particles studied by NMR, in *Physics and Chemistry of Finite Size Systems: From Clusters to Crystals*, P. Jena, B.K. Rao and S.N. Khanna (eds), vol. I. Kluwer Academic, Dordrecht, 1992, 537.
43. Y.Y. Tong and J.J. van der Klink, *J. Phys. Chem.*, 1994, **98**, 11911.
44. Y.Y. Tong, A. Wieckwoski and E. Oldfield, *J. Phys. Chem. B*, 2002 **106**, 2434.
45. W.P. Halperin, *Rev. Modern Phys.*, 1986, **58**, 533.
46. A. Kawabata, *J. Phys. Soc. Jpn.*, 1970, **29**, 902.
47. K. Saiki, T. Fujita, Y. Shimizu, S. Sakoh and N. Wada, *J. Phys. Soc. Jpn.*, 1972, **32**, 447.
48. B. Xu and L. Kevan, *J. Chem. Soc. Faraday Trans.*, 1991, **87**, 2843.
49. R. Scheuermann, E. Roduner, G. Engelhardt, D. Herlach and H.-H. Klauß, *Phys. Rev. B*, 2002, **66**, 144429.
50. J. Michalik, D. Brown, J.S. Yu, M. Danilczuk, Y.J. Kim and L. Kevan, *Phys. Chem. Chem. Phys.*, 2001, **3**, 1705.
51. T. Schmauke, R.-A. Eichel, A. Schweiger and E. Roduner, *Phys. Chem. Chem. Phys.*, 2003, **5**, 3076–3084.
52. H. Hori, T. Teranishi, T. Sasaki, M. Miyake, Y. Yamamoto, S. Yamada, H. Nojiri and M. Motokawa, *Physica B*, 2001, **294–295**, 292.
53. M. Moseler, H. Häkkinen, R.N. Barnett and U. Landmann, *Phys. Rev. Lett.*, 2001, **86**, 2545.
54. T. Teranishi, H. Hori and M. Miyake, *J. Phys. Chem. B*, 1997, **101**, 5774.

CHAPTER 6
Thermodynamics of Finite Size Systems

1 Limitations of Macroscopic Thermodynamics

1.1 A Formal Approach

When we talk about finite size thermodynamics we have to be clear whether we talk about a closed system of a *single nanoscopic ensemble* or about a system that contains a collection of a *large number of nanoscopic ensembles*. In the latter case our observables represent thermodynamic averages over all ensembles. The properties of the individual ensemble get lost in the average, and the validity of several aspects of thermodynamics is restored. The conventional thermodynamic formalism can be adapted to large numbers of small systems by including a size-dependent term in the free energy. It affects properties such as heat capacity, phase transition temperatures, sharpness of phase transitions, *etc.*[1]

Before Gibbs, the basic thermodynamic expression for the internal energy of a macroscopic ensemble at equilibrium was

$$dU = TdS - pdV \tag{1}$$

relating incremental changes of energy, heat and volume work. Gibbs generalised it to account explicitly for variations in the numbers of moles, n_i, to allow for chemical reactions and phase equilibria

$$dU = TdS - pdV + \sum_i \mu_i dn_i \tag{2}$$

where $\mu_i = (\delta U/\delta n_i)_{S,V,nj}$ is called the chemical potential of component i. Extending this philosophy we may add another term $U'dN$

$$dU = TdS - pdV + \sum_i \mu_i dn_i + U'dN \tag{3}$$

where $U' = (\delta U/\delta N)_{S,V,n}$ is kind of a system (rather than molar) chemical potential that vanishes for macroscopic systems where surface and edge effects are negligible, but it is non-zero and size-dependent for small systems. It has been called a 'subdivision potential'.[1] The partition function that includes this subdivision potential has been given by Hill for several sets of variables.[1]

The above treatment is a formal approach that allows the inclusion of surface effects, provided that U' is known. It holds true as long as the basic concepts of thermodynamics remain valid. For sufficiently small systems, however, some of these concepts break down or lose their meaning, so we have to be careful about our language and specify more precisely what the system is that we are looking at. A short description of such pitfalls is given below.

1.2 Systems Beyond the Thermodynamic Limit

We are used to treating macroscopic thermodynamic systems which are ensembles consisting of an infinite number of particles, so that all but the most probable distribution over states can be neglected. These conditions lead to the Fermi–Dirac or Bose–Einstein quantum statistics, and in the classical limit ($kT >> \Delta E$), to Boltzmann statistics. In real systems, the fluctuations of observables are related to the entire ensemble and in general scale as $N^{-1/2}$. As long as N is in the order of N_A members, fluctuations are therefore small. Nanoscale systems consist of a far lower number of ensemble members. For small values of N this leads to a number of situations which require corrections relative to the thermodynamic limit

- *Stirling's approximation* should be used in its more rigorous form which reads: $\ln N! = (N + 1/2) \ln N - N + 1/2 \ln (2\pi)$. It holds to better than 1% down to $N = 4$ and is only 6% off for $N = 2$. The coarser approximation which is often used in statistical thermodynamics, $\ln N! = N \ln N - N$, gives only 48.6% of the true value for $N = 4$, and it is still off by more than 8% for $N = 15$.
- Distributions other than the most probable one assume a non-negligible weight. For small isolated systems the Boltzmann distribution and also Fermi–Dirac and Bose–Einstein statistics are no longer exact in their usual approximations. The true microcanonical energy falls faster than exponential at high energies. Deviations are inversely proportional to the number of particles.[2]
- Computer simulations done with periodic boundary conditions to eliminate surface effects constrain the centre of mass. Therefore, there are only $(N - 1)$ independent particles. In all statistical formulae, N has to be replaced by $(N - 1)$, and the total energy is $(N - 1)$ times the average energy of an individual particle. For the same reason, the equipartition theorem breaks down in a mixture of particles of different mass, so that light particles adopt a higher average kinetic energy than the heavier ones.[2]
- *For an individual ensemble that is decoupled from its environment, thermal fluctuations become increasingly important*, and a time-averaged definition of thermodynamic functions is more useful.[3] The effect has been confirmed by the observation of very large density fluctuations of an ensemble of 10^6–10^7 cyclohexane molecules confined between two mica sheets at a distance of 1–3 nm. Their relaxation evolves over more than 1 min.[4-6]
- A more fundamental effect that is based on the *fluctuation theorem* has to do with time scale as well. The *Second Law of Thermodynamics* teaches us

that a non-equilibrium system always attempts to reach equilibrium, and this process which is called *spontaneous* is irreversible and associated with a positive entropy production. This does not have to be true for finite size systems. Consider, for example, a small system consisting of a number of noble gas atoms A and B which are initially not fully mixed. The Second Law states that the system approaches equilibrium in a continuous process. However, any atom near the dividing surface between the two regions of primarily A and B is free to choose its trajectory, and it is conceivable that at any given time of observation there are many atoms with a trajectory pointing in the direction away from equilibrium. The smaller the system the higher the probability that the number of atoms travelling in the wrong direction exceeds the ones obeying the Second Law. This means that entropy fluctuates, and for short periods of time it will decrease instead of increasing (imitating the 'drunken sailor': two steps forward, one step backward).

Alternatively, instead of observing the trajectories of a large ensemble at a given instance of time we can follow the trajectories of a few atoms of the ensemble for a longer period of time. The *ergodic theorem* states that the two experiments should lead to the same result. At the beginning of the experiment, the number of trajectories which correspond to positive entropy production balances the ones with negative entropy production. The longer we integrate the trajectories of each individual atom, the more we will find that they all tend to point in a direction that is compatible with the Second Law of Thermodynamics. Mathematically speaking, the fluctuation theorem relates the probability P of observing a trajectory of duration t with entropy production $\Sigma_t = X$, to that of a trajectory with the same magnitude of entropy change but where the entropy is consumed:

$$\frac{P(\Sigma_t = X)}{P(\Sigma_t = -X)} = \exp(X) \qquad (4)$$

This means that equilibrium is reached exponentially. An experimental verification of this theorem has been reported recently for the trajectory of a colloidal particle captured in an optical trap that is translated relative to surrounding water molecules.[6] It was found that a significant number of ensemble members violate the Second Law of Thermodynamics for times of 2–3 s! Experimental sensitivity limits the maximum time for which violations can be observed.

• Macroscopic systems exhibit a continuous noise spectrum. Nanosystems, in contrast, are quantum systems. Dissipation is caused by transitions between its discrete energy levels. The noise spectrum therefore has peaks at frequencies corresponding to the differences in energy between these levels. 'Listening' to the intrinsic noise of a system in equilibrium can provide the same spectroscopic information as does probing it by inducing transitions with an external field. This was demonstrated in a study of magnetic fluctuations in rubidium and potassium vapour.[7,8]

- *The local temperature is no longer meaningfully defined* in quantised systems on the nanometre length scale.[9] In a classical system out of equilibrium we are used to finding that energy flows from a hot spot in contact with a colder one in a way that the temperature difference decays exponentially. Between quantised subsystems, energy can be exchanged only in finite quantities so that local 'temperature' fluctuates unpredictably over time. In their work, Hartman *et al.*[9] define a size below which it is no longer meaningful to speak about 'temperature'. They apply their simple model to a string of carbon atoms and conclude that this limit is reached near a length of 10 μm, *i.e.* less than 10^5 atoms. It could mean that this leads to problems when entities such as carbon nanotubes are used as tips of scanning tunnelling microscopes or as building blocks of miniature electronic devices where it may be necessary to reach low temperatures.

1.3   The Breakdown of the Concept of Phases

The above effects are related to statistics. They average out in large ensembles of small systems. Therefore, the following effects are more conceptual rather than gradual, they occur in small systems but do not disappear in large ensembles of the same systems.

- *The concept of phases loses its meaning.*[10] When we ask whether a cluster is metallic, or whether it is liquid or solid, then we should remember that a small cluster may be much like a molecule. Would we ever ask what the melting point of a single small molecule is or whether it is metallic? For large molecules this may make more sense. It is meaningful to discuss the conductivity of a DNA molecule, or to look at its denaturation as a type of melting. Furthermore, we can imagine that one part of a biomolecule denaturates before another one does, and that both forms coexist. In the same way, other than for macroscopic amounts of the same materials, solid-like and liquid-like parts of a cluster may *coexist over some finite temperature range*. The coexistence may be described by an equilibrium constant of the ratio between the numbers of clusters being solid and molten, respectively, in an ensemble. Small clusters switch back and forth between entirely hot solid states with low potential and high kinetic energy, and cold molten equilibrated states with high potential but low kinetic energy. Larger ones approach the behaviour of the bulk with a fraction of a cluster in the solid and simultaneously another in the molten state.

  Figure 1 illustrates this situation by comparing the mol fractions of the acid and base forms near the pK_s of a pH equilibrium, with which we are all familiar, with liquid and solid fractions of nanoscopic amounts of matter near their nominal melting point T_m which is below the bulk melting point T_m*. Just as is the case for an ordinary equilibrium in which the fractions of a solid–liquid equilibrium are related to an equilibrium constant which is given by $K = [\text{liquid}]/[\text{solid}] = \exp(-N\Delta\mu/k_B T)$, where N

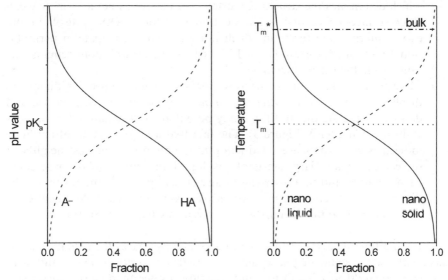

Figure 1 *pH equilibrium (left) in comparison with solid–liquid equilibrium of a nano-cluster (right). While bulk matter has a sharp melting point at $T_m{}^*$ (dashed–dotted line), nanomatter shows phase coexistence over a significant temperature range around a depressed melting point T_m (dotted line). On top of the equilibrium solid–liquid fractions for a given energy (or nominal temperature), there are large fluctuations so that at times everything is solid and at other times everything is liquid*

is the number of members of the ensemble, and $\Delta\mu$ is the free energy difference per member atom or molecule. Thus, for a small ensemble $|N\Delta\mu/k_B T|$ is very small when $\Delta\mu$ changes sign around T_m, so that K does not deviate too much from unity. However, for a bulk ensemble, where N is in the order of N_A, $|N\Delta\mu/k_B T|$ is huge and the equilibrium constant K switches between a negligible and a huge value in an extremely small temperature interval around $T_m{}^*$.[10]

- *The distinction between first- and second-order phase transitions and that between component and phases blur.*[11] Phase transitions are phenomena which are related to cooperativity of a large number of particles. Phase changes of small systems can therefore no longer be classified as being of first or second order according to Ehrenfest. Rather, we have to look for step-like or sigmoid-shaped changes in the mean square displacement of atoms in a certain regime of a cluster. When different phase-like regimes coexist in equilibrium it reminds much more of chemical components or of isomers rather than of phases, and the Gibbs phase rule loses its meaning.
- For small clusters and pore-confined systems there are many more varieties of phase-like forms. All clusters except He_n behave like solids at sufficiently low internal energy ('temperature'), many with well-ordered geometries but often not with symmetries which permit a periodic lattice. At higher temperature, depending on the shape and range of the interaction potential

which determines the structure, clusters can behave as liquids, or they can show surface-melted or less often core-melted forms. Others display more exotic intermediate forms in which not all atoms participate in permutational motions, for example, in Li^{8+}, where the central atom remains the same while the others exchange.[10]

- Molecules confined within narrow pores of a few molecular diameters display a number of *layering transitions*, depending on temperature and density. These are due to three-body potentials or 'oscillating forces' near walls (see Chapter 3, Figure 3). The fact that a large fraction of confined molecules will experience a reduction in the number of nearest-neighbour molecules leads to large shifts of gas–liquid and liquid–solid phase coexistence curves and to a lowering of any critical points.[12] Interestingly, the different layers represent *new fluid phases*, which appear below new associated critical points. Details will be given in Chapter 7, Section 4.

If these different forms of matter are accepted as phases then small systems have rich, size-dependent phase diagrams with *new critical points* and with *coexistence bands* of temperature and pressure, rather than coexistence lines. The transitions between these phase-like forms are not sharp. This is also reflected in sometimes unusual but as yet little investigated heat capacity curves. The latent heats are 'distributed' over some temperature range over which the heat capacities are very high. This occurs also as a consequence of the anharmonicity of intermolecular potentials.[13] On the other hand, heat capacities can become *negative* over a certain temperature range, which seems a sufficient but not a necessary condition for the coexistence of two phases of clusters.[10] Furthermore, the discrete energy spectrum of a system which may often have a small gap above the ground state can lead to a low temperature behaviour that is exponential rather than increasing with T^3 as expected for bulk solids with a Debye-type energy spectrum.[13]

Another important aspect in experiments as well as in molecular dynamics simulations is that of the *time scale of observation and equilibration*.[10] We may take a sequence of snap shots and get the impression that a system is frozen, but if we do the same thing on a longer time scale we will see atoms exchange and details of the structure change, mimicking a liquid. We have to monitor a system for long enough time so that a representative average of mean square displacements can build up.

2 The Basics of Capillarity

The two fundamental equations which describe the main effects of capillarity are ascribed to Young and Thompson (Lord Kelvin). The *Young equation* describes the contact angle θ of a liquid meniscus on a solid substrate as a function of surface tensions. A surface tension relates to the work needed to create a surface, for example cleaving a bulk solid or liquid material. It is defined as a free energy per unit area, or equivalently as a force per unit length

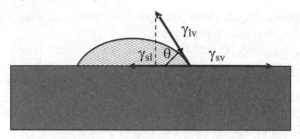

Figure 2 *The contact angle θ of a liquid meniscus on a solid substrate is a result of the equilibrium of the surface tensions between solid and vapour, γ_{sv}, solid and liquid, γ_{sl}, and between liquid and vapour, γ_{lv}*

along a line on the surface. The latter definition leads to an immediate understanding of Young's equation on the basis of the vector diagram shown in Figure 2. For stationary conditions the components of the forces acting along the surface on the solid–liquid–vapour phase boundary line cancel, *i.e.*

$$\gamma_{sv} = \gamma_{sl} + \gamma_{lv} \cos \theta \tag{5}$$

Systems with $90° < \theta \leq 180°$ are called *non-wetting*, those with $0° < \theta \leq 90°$ *wetting* and $\theta = 0°$ relates to the case of formation of a *spreading* liquid which leads to the formation of a liquid film.

The second fundamental relation is the *Kelvin equation*. It describes the size dependence of the vapour pressure of a small free spherical liquid droplet as

$$p(r) = p_0 \exp\left\{\frac{2V_m(l)\gamma_{lv}}{rRT}\right\} \tag{6}$$

where p_0 is the vapour pressure of the infinite size bulk material, γ_{lv} the liquid–vapour surface tension, $V_m(l)$ the molar volume of the liquid, R the gas constant and r the droplet radius. Thus, small droplets are unstable with respect to the bulk. This is plausible since small systems are destabilised by the additional surface free energy $4\pi r^2 \gamma_{lv}$ which does not have to be spent in bulk.

In the more general case a curved surface is not part of a sphere. It can be shown that at each point of a surface the curvature is described by sections of two orthogonal circles (Figure 3a). If one of the circles is chosen in the direction which gives the largest radius, r_1, then the other one has the smallest radius, r_2. This leads to the generalised Kelvin equation which is called the *Young–Laplace equation* for which $r_1 = r_2$ restores immediately (Equation (6)).

$$p = p_0 \exp\left\{\frac{V_m(l)\gamma_{lv}}{RT}\left(\frac{1}{r_1} + \frac{1}{r_2}\right)\right\} \tag{7}$$

It should be noted that r_1 and r_2 can be either positive or negative. The positive sign applies for a convex surface, that is for the droplet shape or a non-wetting liquid in a pore (Figure 3c); whereas, the negative sign applies to the concave case of a wetting surface in a pore (Figure 3b) or the inner surface of a bubble. The fact that the equilibrium vapour pressure is the same everywhere above a given surface leads to the *principle of constant curvature*, i.e. $r_1^{-1} + r_2^{-1}$ is the same at each point of the surface.

Considering Equation (7) with the correct sign of r immediately makes it clear that the vapour pressure of a wetting liquid in a pore is below that of the bulk liquid. As a result of this, a liquid condenses in a pore at a pressure below the bulk vapour pressure, p_0, at the same temperature. Conversely, case (Figure 3c) is realised only for $p > p_0$, which means that a liquid does not spontaneously enter the pore, and so an extra pressure has to be applied.

The situation can be rationalised on the basis of Figure 4 which assumes that a 13-atom cluster is cut out of a bulk close-packed material and leaves behind a void of the same size. The central atom of the cluster (black) is coordinated by 12 atoms. This corresponds to the full coordination in the bulk. Each surface atom of the cluster is in contact with five neighbouring cluster atoms, giving a coordination number of five. The situation of the inner surface of the void is more complicated. It can be visualised by adding atoms to the cluster surface in the regular positions of the extended bulk structure. These adatoms can be in

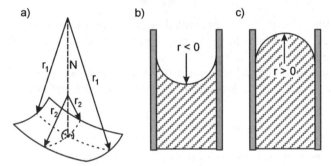

Figure 3   *The curvature at point P of a general surface is characterised by a maximum radius r_1 and a minimum radius r_2 of two orthogonal circles (a). A wetting liquid in a pore has a concave surface with $r < 0$ (b), a non-wetting liquid has a convex surface corresponding to $r > 0$ (c)*

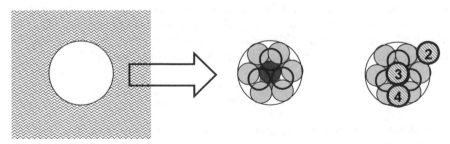

Figure 4   *A 13-atom close-packed cluster cut out of a bulk material, leaving behind a void. On the right, three adatoms (hashed) with coordination numbers as indicated are added to the cluster surface to visualise the options for the atoms at the inner surface of the void*

contact either with four, three or two cluster atoms (right side of Figure 4), so that the eight, nine or ten atoms complementing the full coordination must be at the inner surface. We see that atoms at the void surface have a higher coordination number and thus a higher stability than the ones at the outer surface, reflecting directly the lower vapour pressure in the void compared to the gas phase in equilibrium with a spherical cluster. By comparison, each of the atoms of a planar single crystal (111) surface has a coordination number of nine.

A second point which derives from Figure 4 relates to the surface tension of a curved surface. When the cluster is cut out, bonds (no matter whether they are of covalent, ionic or van der Waals type) are broken at each point of contact, and a dangling bond is left behind on both the cluster surface and the inner surface of the void. Due to the symmetry of this situation, the surface free energy or surface tension is to first order the same for the two surfaces, independent of their curvatures It is strictly true only as long as pair potentials are additive and where long-range forces and many-body effects are negligible. While this holds well for van der Waals substances it does not apply to metallic, ionic or hydrogen-bonded systems.[14] It is also plausible that convex and concave surfaces undergo surface reconstruction or relax differently, so that eventually the surface tension may often depend somewhat on the sign and extent of curvature. Not much is known about this in great detail, and we shall in the following mostly assume a surface tension independent of curvature.

For a pore-confined liquid near the critical point the Kelvin equation is written more generally.

$$p = p_0 \exp\left\{ \frac{V_m(l)V_m(v)}{V_m(v) - V_m(l)} \times \frac{\gamma_{lv}}{RT} \left(\frac{1}{r_1} + \frac{1}{r_2} \right) \right\} \qquad (8)$$

Far from the critical point $V_m(l)$ can be neglected against $V_m(v)$, and Equation (8) reduces to (7), but the more we approach this point the more important it is to use the full expression. It should furthermore be noted that near the critical point the distinction between liquid and vapour gets lost and γ_{lv} approaches zero.

In the above equation, the effective pore radii r_1 and r_2 decrease during the progress of adsorption, leading to a simultaneous decrease of p. A different approach to account for the adsorbed multilayer film of increasing thickness $t(p)$ which is formed prior to pore condensation is given by the Kelvin equation as

$$p = p_0 \exp\left\{ \frac{2V_m(l)\gamma_{lv}}{RT} \left(\frac{1}{r_0 - t(p)} \right) \right\} \qquad (9)$$

Thereby, the analytical dependence of $t(p)$ is often assumed to be based on the process of formation of a liquid-like film on a flat surface as described by the *Frenkel–Halsey–Hill theory.*[15] It relates the film thickness t to the pressure p as

$$\ln\left(\frac{p}{p_0} \right) = -\alpha t^{-m} \qquad (10)$$

where α is the fluid-wall interaction parameter and m which has a theoretical value of 3 for van der Waals fluids is experimentally often found to be in the range 2.5–2.7.[15]

The Kelvin equation should be valid for large pores well below the critical temperature. Its derivation is based on several assumptions: density oscillations near the wall (see Chapter 7, Section 4.2) should be negligible, the system has to be large enough so that a surface tension can be defined, the gas phase is approximated as an ideal gas, the liquid phase is incompressible and the surface tension can be taken as the value for the bulk fluid at this temperature.[12]

3   Phase Transitions of Free Liquid Droplets

As long as the concept of phases is meaningful we can talk about phase transition temperatures. The shifts of freezing and boiling points of simple liquids as isolated droplets or clusters, or confined in cylindrical or slit-shaped pores have been well known for some time. The coexistence of two phases is given by the *Clausius–Clapeyron equation*, which for liquid–vapour coexistence in a range between any reference temperature T_0 and the temperature T of interest is expressed as

$$p(T) = p(T_0) \exp\left\{ -\frac{\Delta H_{vap}(T_0 - T)}{RTT_0} \right\} \tag{11}$$

Here, ΔH_{vap} is the molar latent heat of evaporation. At the same time, the vapour pressure on a curved surface obeys the Kelvin equation. At the phase transition, p_0 of Equation (6) is given by Clausius–Clapeyron so that

$$p(r) = p(T_0) \exp\left\{ -\frac{\Delta H_{vap}(T_0 - T)}{RTT_0} \right\} \exp\left\{ \frac{2V_m(l)\gamma_{lv}}{rRT} \right\} \tag{12}$$

and $p(r)=p(T_0)$. The product of the two exponential functions is therefore equal unity, or

$$\frac{\Delta H_{vap}(T_0 - T)}{RTT_0} = \frac{2V_m(l)\gamma_{lv}}{rRT} \tag{13}$$

which rearranges to

$$\frac{T_b - T_b^*}{T_b^*} = \frac{\Delta T_b}{T_b^*} = -\frac{2V_m(l)\gamma_{lv}}{\Delta H_{vap}r} \tag{14}$$

where T_0 now signifies the boiling point of the bulk system, $T_b^*(r = \infty)$, and $T = T_b$. It means that the increase of the vapour pressure of a spherical droplet results in a decrease of its boiling point T_b; *i.e.* at a given vapour pressure the droplet boils at a lower temperature compared with the bulk liquid. On the other hand, for a wetting pore confined liquid with a meniscus which is represented by a negative value of r (Figure 3b) the vapour pressure is decreased compared with its bulk value at the same temperature. Therefore, for a given pressure the boiling point increases in the pore with respect to the bulk, or at a given temperature, condensation/boiling occurs at a lower vapour pressure.

In an analogous way, the *Gibbs–Thomson* equation is derived, which describes the depression of the melting point T_m of a free cluster (with melting enthalpy ΔH_m) which is covered by a liquid layer:[16]

$$\frac{T_m - T_m^*}{T_m^*} = \frac{\Delta T_m}{T_m^*} = -\frac{2V_m(l)\gamma_{sl}}{\Delta H_m r} \qquad (15)$$

All of the above formulae assume that the molar volumes, the latent heats of melting and evaporation, and the surface or interfacial tensions assume their bulk values and are size-independent. This is in fact not the case. We are already aware that the lattice constants normally shrink slightly as the particles get smaller. This will affect the molar volume by up to a few per cent, which is not normally of great concern. There have been discussions about the size-dependence of phase transition enthalpies, and we shall defer details to Chapter 8.1, but basically, the result is the following: when *all* atoms or molecules of a cluster or pore-confined state are included then the melting enthalpies decrease with decreasing size as the surface atoms have a lower coordination number, and the surface layer may already have melted. However, when a surface layer of suitable thickness is excluded and only the behaviour of the core is discussed, as is often done, there is no significant dependence. The size dependence of the surface or interfacial tension is often neglected as well, yet it is likely that this is also not always a good approximation. Discussions may be found in work by Sinanoglu[17] Jiang *et al.*[18] and Weissmüller.[19]

4   The Lotus Effect

The leaves of the lotus plant (Figure 5, left) always appear to be clean even in an environment where other surfaces are immediately contaminated with dust and dirt particles. They possess a remarkable self-cleaning ability which made the sacred lotus plant a symbol of purity in Asian religions. This phenomenon which was termed the *Lotus effect* was not well understood until recently. As we will show below, it is based mainly on the existence of a micro structured and extremely hydrophobic surface. It is of course attractive to try and mimic the effect for man-made surfaces (right part of Figure 5).

All primary parts of plants except their roots are covered by a cuticle which is composed of soluble lipids embedded in a polyester matrix. Due to its chemical composition the cuticle forms in most cases a hydrophobic interface to its environment. Nevertheless, the static contact angle of water shows a remarkable variation from 28° on the smooth surface of the leaves of rainforest herb (*Heliconia densiflora*), 89° for the evergreen leaf of *Magnolia denudata* to 160° for the rough

(a)   (b)

Figure 5 *Left: Flowers and leaves of the sacred lotus plant (Nelumbo nucifera). Right: Water droplets on a wood surface treated with BASF's lotus spray which makes the surface extremely water repellent (superhydrophobic) and self-cleaning. The contact angle exceeds 140°*
(From Ref. 21. Photo: BASF)

surfaces of the leaves of kohlrabi (*Brassica oleracea*) and the celebrated lotus leaf (*Nelumbo nucifera*). It was shown that there is interdependence between the water repellent nature, the reduced adhesion of dirt and dust particles and the surface roughness of these leaves.[20] Scanning electron microscopy of leaf surfaces revealed a rich microstructural diversity. Cuticular folds and epicuticular wax crystalloids cover the cuticular surface (Figure 6) and form a microrelief of about 1–5 μm height, which is a key element for water repellency.

The increased contact angle with concomitant increasing hydrophobicity is understood on the basis of three cumulative mechanisms, relating to (i) surface roughness, (ii) entrapped air in the structured surface underneath the liquid, and (iii) surface coverage with hydrophilic hair.

(i) The effect of surface coverage is described by *Wenzel's law*[22] which relates the effective contact angle θ' of a rough surface to its intrinsic value θ *via*

$$\cos \theta' = r \cos \theta \qquad (16)$$

In this equation r is the *surface roughness*, the ratio of the real surface area divided by the projected flat surface area, a number that is always greater than 1. It holds true as long as there is no air entrapped in the pits and grooves of the structured surface. From electron microscopy images of the cuticular wax crystals of Indian Cress (*Tropaeolum majus*) leaves an approximate value of $r \cong 4.7$ was estimated.[23,24] This is sufficient for an increase of the intrinsic $\theta \cong 100°$ of the cuticular wax to $\theta' \cong 145°$, but not sufficient to explain the observed contact angle which is close to 180°.

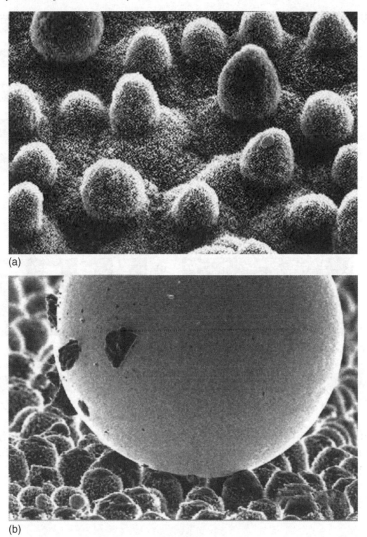

(a)

(b)

Figure 6 *Scanning electron micrographs of the leaf surface of a lotus plant, Nelumbo nucifera (a), and mercury droplets sitting on the surface of taro, Colocasia esculenta, like a fakir on a bed of nails, demonstrating the Lotus effect (b). Contaminating particles adhere to the surface of the droplet and are removed from the leaf while the droplet rolls off (the total width is ca. 200 μm)*
(Reproduced from Ref. 20. Copyright (1997) with permission from Springer)

(ii) A more detailed analysis of the water drop on Indian Cress at grazing incidence reveals a shiny layer which is due to total reflection of the light by small air compartments at the interface. In this case, Equation (16) is no longer valid. To describe the situation which is sketched schematically in Figure 7, Young's equation for a simple, smooth surface (Equation 5) is rewritten to

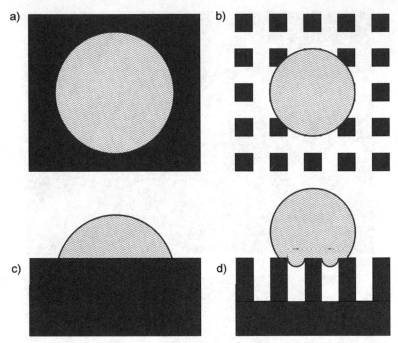

Figure 7   *Schematic top view (a,b) and side view (c,d) of the effect of a composite or microstructured surface on the contact angle of a liquid droplet. The left side shows a drop that wets the dark substrate. The right-hand side reflects the situation when the same substrate covers only ca. 30% of the surface and is interdispersed by a non-wetting surface or by air*

account for a composite surface with a fraction f of one substrate and a fraction $(1-f)$ of a second substrate (denoted by an asterisk superscript). The surface free energy γ_{sl} is then given by

$$\gamma_{sl}' = f\gamma_{sl} + (1-f)\gamma_{sl}^* = \gamma_{sv} - \gamma_{lv}\cos\theta' \tag{17}$$

Solving for $\cos\theta'$ yields

$$\cos\theta' = \frac{f(\gamma_{sv} - \gamma_{sl}) + (1-f)(\gamma_{sv} - \gamma_{sl}^*)}{\gamma_{lv}} = f\cos\theta + (1-f)\cos\theta^* \tag{18}$$

This means that $\cos\theta'$ is a weighted average over the two substrates, the weights being given by the surface fractions. For the case where the second substrate is the vapour that is trapped between the microstructural posts (*i.e.* $\gamma_{sl}^* = \gamma_{lv}$) we obtain

$$\cos\theta' = f\cos\theta + (1-f)\left(\frac{\gamma_{sv}}{\gamma_{lv}} - 1\right) \tag{19}$$

which for a hydrophobic substrate (*i.e.* $\gamma_{sv} << \gamma_{lv}$) converts to

$$\cos\theta' = f(\cos\theta + 1) - 1 \qquad (20)$$

The apparent contact angle according to Equation (20) is displayed in Figure 8 for water on Teflon consisting of microposts which cover a fraction f of the entire surface (compare with Figure 7d). We note that a remarkable apparent contact angle above 175° is obtained when the fractional contact area of water on Teflon is below 1%.[25] On lotus leafs the surface fraction that is in contact with water amounts to 2–3%.

(iii) We still owe an explanation for the hydrophobicity of hairy surfaces, as for example that of Lady's Mantle (*Alchemilla vulgaris*) leaves.[23] The interesting point here is that the hairs are more hydrophilic than the cuticular surface, exhibiting a contact angle of less than 60°. This has the consequence that small droplets nucleating on the cuticula are lifted off into the hairy brush and roll up to near-spherical shape to lower the total energy as soon as they make contact with the hairs. In part, the surface energy gained is spent as elastic energy when a number of hairs are bent into a bundle.[23]

A small contact area leads to a low energy of adhesion. On such a surface, a water drop gains very little energy through adsorption. The driving force for distortion of a drop from its natural spherical shape is therefore small. The contact angle of the drop depends almost entirely on the surface tension of the water and very little on that of the surface.

A low energy of adhesion also means low friction for the drop to move on the surface. On highly hydrophobic surfaces, the drops run off at a very low angle of inclination. In fact, on an artificially nanostructured surface consisting of an array of sharp Teflon-coated tips the droplets could not be stabilised and ran off at an inclination angle far below 1°.[25] This reduced friction applies not only to an open surface but to a similar extent also to the movement of a drop in a channel. It is crucial in any case that the structure is on a micrometre to

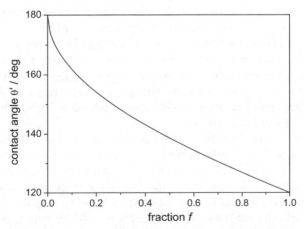

Figure 8 *Apparent contact angle of water on a microstructure consisting of Teflon posts which cover a fraction f of the total surface (Equation 20)*

nanometre scale, since otherwise the liquid would lose levitation under slight pressure.

It is important to note that a real surface is often structured on different length scales, as seen well in Figure 6 (upper part). This leads to the concept of a hierarchically rough surface which under certain conditions will be superhydrophobic.[24]

What is now the secret behind the *self-cleaning property* of a lotus leaf? The important point is that because of the reduced contact area, adhesion of any dust particle on a microstructured surface is also very low. Whenever a water drop comes along, it is energetically favourable for a particle to adhere to the water surface (just as to the mercury surface in Figure 6). It sticks to it when the droplet runs off and is removed from the surface in this way. In contrast, especially hydrophobic particles tend to remain on a flat surface. The water drop is not able to pick them up and serves merely to push them around and redistribute them on the surface.

Why should a plant design a strategy to fight dirt? Well, it may not so much be the normal dirt that worries the plant, but wettability is also important for the adhesion of microorganisms and their spores and conidia to leaf surfaces. Therefore, the epicuticular wax crystalloids and their physical properties may be regarded as the first line of defence against pathogens. The Lotus effect is also not restricted to plants. It is equally important for insects, especially those with large wings which cannot be cleaned by legs. Water repelling wing surfaces exhibit self-cleaning ability. This allows for the maintenance the flight capability as it avoids unequal and extra load on the wings.

The Lotus effect also has important technical implications. Water repellent fabrics for sports clothing already work in principle on the same basis. The effect of the hydrophobic fabric fibres is enhanced by the reduced contact area with open space for transpiration in between. Microfibre fabrics are used for furniture cushion because of their dirt repelling properties. Microstructured surface treatments are attractive for glasses and for all places where cleanness is crucial or where vigorous rubbing for cleaning may cause damage of fragile structures. The effect should also prove extremely important for microfluidic devices. Classical flow of a Newtonian fluid in cylindrical tubes according to the Hagen–Poiseuille law scales with the fourth power of the tube radius, it is thus extremely unfavourable in narrow tubes. In Newtonian flow the first liquid layer sticks to the surface. In contrast, on a microstructured hydrophobic surface the first layer can slide, which should permit a higher throughput of liquid under the same conditions.[25]

A truly spectacular effect of a smart surface that can switch from superhydrophobic to superhydrophilic behaviour under UV-illumination was reported recently by Feng *et al.*.[26] The effect is based on a film of upright ZnO nanorods (Figure 9) which provide the posts that enable a fractional coverage of the surface, as discussed above. In combination with the hydrophobic character of ZnO in the dark, this leads to a contact angle of 160° for water. A preliminary explanation of the effect is as follows: UV-illumination is thought to create electron-hole pairs which segregate on the surface. As a result, the surface

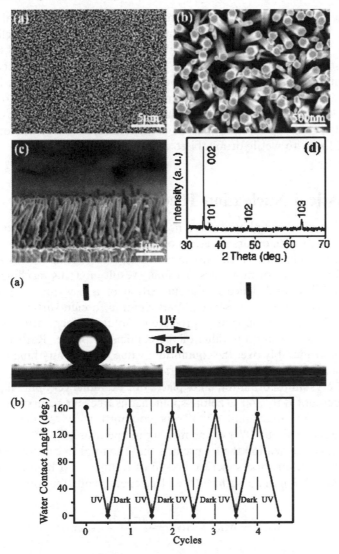

Figure 9 *SEM images of ZnO nanorod films at low and high magnifications, cross-sectional view of aligned ZnO nanorod film and XRD pattern (a). Photographs of water droplet shape before and after UV illumination, along with corresponding evolution of contact angles (b)*
(Reprinted with permission from Ref. 26. Copyright (2004) American Chemical Society. Image courtesy L. Jiang)

becomes hydrophilic, and the water drop expands to a smooth film with an effective contact angle of 0° on the relatively flat surface.

A related hydrophilic/hydrophobic switching effect on the basis of temperature change between 30°C and 40°C was reported by the same group for groove-structured silicon surfaces modified with poly(N-isopropylacrylamide).[27]

Possible future applications of a spray that is under development at BASF (compare Figure 5) include leather impregnation and bringing self-cleaning shoes into sight. However, several problems remain to be solved. At this point, the nanostructures are somewhat too large, so that a film is not completely transparent, rendering a surface opaque and dull. Furthermore, the layers rub off or scratch easily. The most difficult problem to overcome is that oils tend to migrate into the nanostructures, which renders them ineffective.[21] Of course, surfactant-containing solutions have a greatly reduced surface tension and are, therefore, likely to wet hydrophobised surfaces also.

5 Classical Nucleation Theory

The creation of a new phase from a homogeneous metastable (supersaturated or supercooled) state occurs *via germs* or embryos of the new phase. Germs are small transient clusters which exist in constantly fluctuating size by desorbing and attaching atoms or molecules. This may result in germs reaching a critical size for stability, and hence in the nucleation of a new phase. The normal observation is that in the absence of participating foreign surfaces, nucleation does not take place if the vapour pressure or solute concentration is only just slightly over the saturation value of the condensed phase. Rather, it can be increased considerably over the equilibrium value without anything happening, until, at some fairly sharp limit, general condensation in the form of fog or droplet or crystallite formation takes place. For example, very pure liquid water can be supercooled to $-40°C$ before spontaneous freezing occurs. An empirical rule says that the maximum ΔT of supercooling for crystallisation from the melt amounts to about 18% of the melting temperature, T_m.[28]

The classical nucleation theory predicts that the total free energy cost, ΔG_{tot}, of forming a nucleus of a spherical crystal of radius r in a liquid consists of a bulk volume (ΔG_v) and a surface (ΔG_s) free energy contribution,

$$\Delta G_{tot} = \Delta G_v + \Delta G_s = \frac{4\pi r^3 \rho \, \Delta\mu}{3} + 4\pi r^2 \gamma \qquad (21)$$

where γ is the interfacial free energy, ρ is the density of the bulk liquid and $\Delta\mu$ is the chemical potential difference between the bulk solid and the bulk liquid.[29] Analogous expressions hold for nucleation of a solid or a liquid droplet from the vapour phase, with $\Delta\mu = -RT \ln x$, where $x = p/p_0$ is the actual vapour pressure relative to its saturation value, or for crystallisation of a non-electrolyte from solution where $x = S/S_0$ is the concentration in units for the saturation concentration.

The volume term in Equation (21) scales as r^3 and the surface term only as r^2. For sufficiently small nuclei ΔG_{tot} is therefore dominated by the surface term, which leads to an activation barrier ΔG^* for nucleation (Figure 10b). At the phase transition point the two phases are in equilibrium, and $\Delta\mu$ equals zero. Below this point the system is supersaturated so that $\Delta\mu$ is negative.

Supersaturation increases in magnitude as the temperature offset increases (Figure 10a). It is thus the supersaturation which drives crystallisation.

It is straightforward to obtain from Equation (21) the activation barrier ΔG^* at the critical nucleation radius $r_c = -2\gamma/\Delta\mu$:

$$\Delta G^*(r_c) = \frac{16\pi\gamma^3}{3(\Delta\mu)^2} = \frac{4\pi r_c^2 \gamma}{3} \tag{22}$$

This is one-third of the surface free energy of the whole nucleus. The critical radius decreases with increasing supersaturation, *i.e.* with $\Delta\mu$, and the activation energy increases with the inverse square of $\Delta\mu$.

Experiments are often performed by changing the temperature rather than the vapour pressure. For this case, Equation (22) is easily converted using the Gibbs–Thomson Equation (15) with $r = r_c$. For condensation with a normal boiling temperature T_b^* this results in

$$\Delta G^*(r_c) = \frac{16\pi\gamma^3}{3} \left(\frac{V_m(l)\, T_b^*}{\Delta H_{vap}\, \Delta T_b} \right)^2 \tag{23}$$

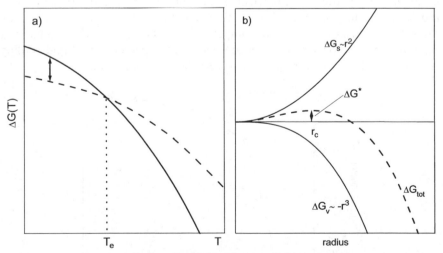

Figure 10 *(a) Course of the molar Gibbs free energy for the high temperature phase (full line) and the low temperature phase (dashed line) near the equilibrium temperature T_e. The lower curve gives the equilibrium state, and the arrow reflects $\Delta\mu$ due to supercooling, the driving potential for crystallisation. (b) Surface and volume contribution to the total free energy change as a function of radius on crystallisation. ΔG_{tot} goes through a maximum at a critical radius r_c*

and

$$\Delta\mu = -RT \ln\left(\frac{p}{p_0}\right) = \frac{\Delta H_{\text{vap}}}{V_m(l) T_b^*} \Delta T_b \tag{24}$$

In analogy to expressions in reaction rate theory the steady-state nucleation rate is given by

$$J = J_0 \exp\frac{-\Delta G^*}{RT} \tag{25}$$

where J is the number of critical clusters formed per unit time in a unit volume of the ambient phase, and J_0 is roughly the collision frequency (collisions per cubic centimetres and seconds) for condensation from the gas phase. J is an extremely strong function of the degree of supersaturation, p/p_0. Thus, increasing the vapour pressure of water at 0°C from $1.5p_0$ to $2p_0$ increases the nucleation rate by 200 (!) orders of magnitude, and another 81 orders of magnitude for an increase from $2p_0$ to $4p_0$ (Figure 11).

For crystallisation from liquid solution at a concentration of n particles per cm³ J_0 is impeded by the free energy of activation of diffusion, ΔG_D, and is given by[28]

$$J_0 = \frac{nkT}{h} \exp\left(-\frac{\Delta G_D}{kT}\right) \tag{26}$$

This means that the degree of supercooling reduces ΔG^* from an expression analogous to Equation (23), in which T_b is replaced by T_m and ΔH_{vap} by ΔH_m, and it speeds up the rate of nucleation. On the other hand the increasing viscosity diminishes J_0 so that competition of the two trends results in a maximum nucleation rate at a certain degree of supercooling. Rapid cooling therefore results in the formation of glasses instead of crystals.

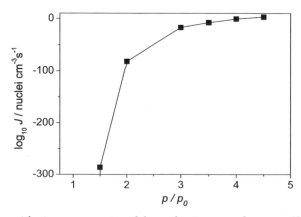

Figure 11   *Logarithmic representation of the nucleation rate of water at 0°C as a function of the degree of supersaturation*
(Figure based on values from Ref. 28)

Classical nucleation theory has important implications for the *kinetic control of crystal growth* in nanotechnology. Once nuclei form in a supersaturated solution they begin to grow by accretion. As a result, the concentration of the solution drops below the critical threshold so that no further nucleation occurs. Existing nuclei grow further until all material is used up. This leads to a relatively monodisperse distribution of crystallite sizes. Thus, if one wishes to grow large single crystals one wants to have only a few nuclei, thus the degree of supersaturation should be small – or even better: supersaturation should be held at a level where nucleation is unlikely, and a few nuclei can be brought in as seeds and grown to the desired size. In contrast, if one wants to have a large number of very small nanoparticles, one should attempt to reach a high initial degree of supersaturation very fast. This latter case is expressed qualitatively by *Weimarn's law*:

$$\frac{1}{d} = \frac{kS}{S_0} \tag{27}$$

where k is some system-specific constant and S/S_0 is the initial concentration in units of the saturation concentration. It tells that the final crystal size d is inversely proportional to the initial degree of supersaturation.

If one wishes to have monodisperse colloidal systems in which the particles have the same size to within 5%, it is often advisable to conduct the synthesis in two steps. The first step consists of fast nucleation. The primary particles are centrifuged off, and in a second step they are grown to the desired form under conditions which do not permit further nucleation. An instructive example of the precision of monodispersity is given in Figure 12.[30]

Artificial rainmaking is an interesting and important example of seeding. At −4°C and a sufficient level of supersaturation the nucleation of water can be enforced by seeding with silver iodide, whose crystal structure is the same as that of ice and cell dimensions are very close to it.[28] On other occasions, it may

(a)   (b)

Figure 12   *Monodisperse spherical colloidal TiO$_2$ particles with a precision of <5% of the size distribution (left), and crystalline order of silica colloids (right)* (Reprinted with permission from Ref. 30. Images courtesy S. Eiden)

be of interest to avoid nucleation and form a glass. This may often be achieved by using a mixed solvent system where both components solvate the solute well in the liquid state, but do not form mixed crystals (*e.g.* water and glycerol).[28]

Another important principle, called *Ostwald ripening*, is derived from the Kelvin Equation (6), which can also be written for the concentration $S(r)$ which is in equilibrium with a crystal of average radius r and surface area averaged surface tension γ_{sl}, as

$$S(r) = S_0 \exp \left(\frac{2V_m(s)\,\gamma_{sl}}{rRT} \right) \tag{28}$$

where S_0 is the saturation concentration above a macroscopic crystal (in the limit of $r \rightarrow \infty$). It tells that $S > S_0$ for all finite size crystals. Therefore, when a batch of crystals is kept in its solution at a concentration just above S_0, the large crystals will grow at the expense of the small ones. This process can be accelerated when the condition is met at a higher temperature where the kinetics of the exchange at the crystal surface is enhanced; but it should be taken into account that for entropic reasons the equilibrium concentration of defects in the crystal is also increased. Furthermore, it should be noted that Ostwald ripening causes large crystals to grow and small ones to shrink. It therefore leads to a broadening of crystal size distribution and is not a suitable method for obtaining small monodisperse particles.

More accurately, according to *Wulff's law* for a crystal in equilibrium with its environment, γ_{sl}/r in Equation (28) should be replaced by the ratio of the surface free energy γ_i of the i-th crystal face divided by its distance measured by its normal from the crystal centre, h_i.[31] Wulff's law states that this ratio is constant for all crystal faces:

$$\frac{\gamma_i}{h_i} = \text{const} \tag{29}$$

It is a result of minimisation of the crystals surface free energy, which requires that high energy surfaces contribute with a smaller fraction to the whole surface whereas the lowest energy surfaces contribute with the largest fraction. For needle-like crystals this means that the high energy surfaces are the small area faces near the tip of the crystallite. Wulff's law holds for crystallites which are not too small so that free energy contributions of edges and corners can be neglected (Figure 13).

An effect related to Ostwald ripening becomes obvious for solid electrochemical systems when Equation (28) is rewritten in terms of the chemical potential, $\Delta\mu = -RT \ln (S/S_0)$, or

$$\mu = \mu(r = \infty) + \frac{2\gamma V_m}{r} \tag{30}$$

Equation (30) means that the surface term leads to an electrochemical force between nanocrystals and bulk material that is separated by an electrolyte. This

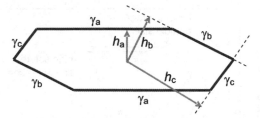

Figure 13 *Illustration of Wulff's law, stating that g_i/h_i is constant for any crystal face in equilibrium with its environment ($\gamma_a < \gamma_b < \gamma_c$).*

has been observed for nanocrystalline copper and silver separated by a Cu^+ or Ag^+ electrolyte from the corresponding bulk metal.[32]

Another effect that is often modelled in terms of Ostwald ripening is *sintering* of metal particles. It is important to understand the kinetics of sintering in order to estimate the lifetime of supported metal nanocatalysts. However, it was found that the energy of a metal atom in a nanoparticle increases much more dramatically with decreasing size than predicted on the basis of the Gibbs–Thomson equation with size-independent surface tension, and that they do not properly describe sintering.[33] It was suggested that a model based on pair-wise bond additivity provides a better description at small particle radii.

The critical radius r_c of nucleation depends on the free energy of activation (Equation 22). It is therefore concentration dependent through the entropy contribution to ΔG^*. A slightly different critical size relates to the growth rate of a preformed particle. Particles below this critical size have a negative growth rate and shrink; those above the critical size grow. When the average size is around the critical size this will cause the small particles to shrink and the larger ones to grow (Ostwald ripening). This results in a broadening of size distribution. Particle growth is normally diffusion controlled. Its consequence had been studied more than half a century ago by Reiss.[34] It leads to an important principle for the growth of monodisperse particle size distributions, as illustrated in Figure 14 which displays the diffusion limited growth rate as a function of particle size for two different solute concentrations.[35] Above this critical size all particles grow. However, it has been shown that the growth rate goes through a maximum (indicated by an asterisk in Figure 14). Below this maximum, the smaller particles grow more slowly than the larger ones, which still results in a defocusing of the size distribution (lower curve). However, above the maximum the smaller ones grow somewhat faster so that we now have a *size-distribution focusing*. The effect has been demonstrated experimentally.[36]

6 Shape Control of Nanocrystals

When a crystal grows each new building unit has the choice of adding to one or another crystal face. Its binding energy is highest when it adds to a high energy surface. This is the reason for the rapid growth of the crystal in a direction

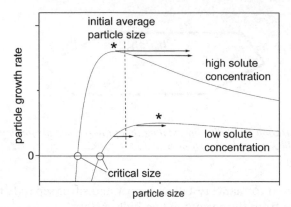

Figure 14 *Particle growth rate as a function of particle size, illustrating the size-distribution focusing (upper curve) and defocusing (lower curve) effect. The asterisk marks the maximum of the diffusion limited growth rate* (Redrawn from Ref. 35 with permission from the publisher)

perpendicular to a high energy surface and a lower growth rate perpendicular to low energy surfaces. It is the kinetic basis of Wulff's law for a crystal shape in thermodynamic equilibrium. On this basis it would appear that there is a single shape of a crystalline material with a given structure of the unit cell. However, under conditions of practical relevance many crystals do not grow in their equilibrium shape. The reason is that the surface free energy of a crystal can be varied without changing the crystal structure, for example by surface melting or by adsorption of various foreign species, and this can dramatically and selectively influence the growth kinetics of crystal faces.

The most familiar example for crystals which can adopt an amazing variation of shapes from needles over plates to most beautiful star-like structures is the snow flake (Figure 15). There is no doubt that the crystal structure of all snow flakes is normal hexagonal ice which has two types of surfaces, the two hexagonal (0001) basal faces perpendicular to the hexagonal symmetry axis (the end faces of a non-sharpened pencil) and the six rectangular prism faces (the colour painted faces of the pencil, *e.g.* (1010), see also Figure 16).

How does it come then that snow flakes can grow in different morphologies, which is seemingly in contradiction to Wulff's law? The answer lies in the changing nature of these surfaces as temperature changes. The situation is illustrated in Figure 17, the so-called *Nakaya diagram*. Down to a temperature of $-4°C$ both types of surfaces are covered with a thin *quasi-liquid layer* (surface melted ice). Under these conditions the growth rate R_{0001} is smaller than R_{1010}, leading to plate-like structures.[37] From $-4°C$ to $-10°C$ the surface of the (0001) plane is no longer liquid, rendering the growth rate along the hexagonal axis larger than the other one, so that long needles or prisms grow. Between $-10°C$ and $-20°C$ both types of crystal faces dry up, but the (0001) planes are smooth and the other ones rough so that the growth rate along the hexagonal axis can no longer compete with the one perpendicular to it, and we

Figure 15 *Selected choice of structures of snow flakes. Note the hexagonal symmetry, independent of whether the flake is a thin plate or a collection of six simple or branched dendrites*
(Reprinted with permission from Ref. 37 by the publisher)

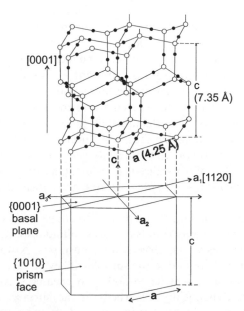

Figure 16 *Crystal structure of ice*
(Reprinted with permission from Ref. 37 by the publisher)

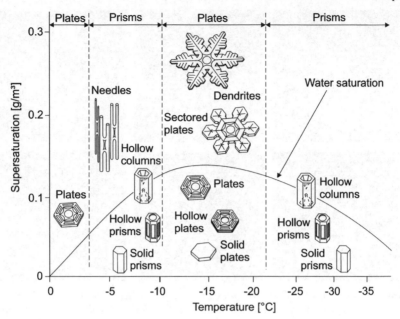

Figure 17 *Morphology diagram of snow flakes due to variation of the growth rates of the*
two crystal faces with changing environmental conditions (Nakaya diagram)
(Reprinted with permission from Ref. 37 by the publisher)

have formation of plates again. Only below $-20°C$ when all crystal faces are smooth do we obtain another regime of formation of needles and prisms (Figure 17). The shapes also depend on the degree of supersaturation. Higher humidity leads to more complex shapes.

A related diagram was reported for the dependence of crystal morphology on the degree of supersaturation and ion strength for $BaSO_4$.[38]

There are numerous examples where humans have been able to control the shape of crystallites by changing the environmental conditions of growth, for example by adding salts or surfactants to the solution, or by changing the electrochemical potential in synthesis by electrodeposition. Other methods involve vapour deposition or template synthesis. This is more than just a game, as electronic properties of delocalised states in metals or semiconductors clearly depend on the dimensions (Chapter 4), and magnetism is also a function of shape (Chapter 5).

The most common method that can lead to shape effects in the synthesis of free-standing inorganic nanocrystals involves colloidal solutions in the presence of surfactant molecules which serve several purposes. At low temperature the surfactants are strongly bound to the surfaces. They passivate them chemically, prevent aggregation and provide solubility in organic solvents. At high temperature they are dynamically adsorbed on the surface, mobile enough to provide access for the addition of monomer units to the growing nanocrystal.[39] This organic coating allows for great flexibility in that the surfactants can be exchanged with different organic molecules with various functional groups. Furthermore, the surfactants can be removed temporarily to permit the growth of an epitaxial layer of another material to produce core–shell particles in which the shell assists further in tuning the electronic, optical or magnetic properties. Furthermore, spherical and non-spherical quantum well structures can be designed.

Shape control has been investigated in detail for the case of the synthesis of CdSe nanoparticles.[39] The traditional method of growing near-spherical particles of this hexagonal (wurtzite) structure is by injecting precursors into a liquid surfactant (90% trioctylphosphine oxide, TOPO) that undergo pyrolysis at temperatures above 300°C. Using a combination of two pure surfactants, TOPO and hexylphosphonic acid (HPA) in various ratios allows a control of nanocrystal shape: a molar fraction of 8% HPA yields faceted 'spheres', 20% HPA gives nanorods, and upon a further increase of the fraction of HPA three-dimensional structures such as pencil-, arrow- and tree-shaped crystals are obtained. The system of this behaviour was analysed in order to understand the mechanism. It is believed that the phosphonic acid is responsible for raising the energy of some of the facets relative to others. It is known that they bind quite strongly to several cations, such as Cd^{2+}. For example, both the (001) and the (00–1) faces can be terminated by either a layer of Cd^{2+} or Se^{2-} ions; however, due to the lack of inversion symmetry in the wurtzite structure the two must be different. It is believed that HPA binds very weakly to the anion-terminated surface but rather strongly to the Cd-terminated one. In this way, the lack of inversion symmetry is mirrored into different growth rates, and together with the fact that the side faces can be passivated by either TOPO or HPA, this is imagined to lead to the special crystal shapes.

Since the CdSe nanocrystals are in the quantum confinement regime of excitons, the band gap is a strong function of crystal size (Chapter 4, Section 5.1). Changing the aspect ratio of a crystal raises some energy levels and lowers others. Pseudopotential calculations on this system predict that there is a crossover of the two highest occupied levels at an aspect ratio of *ca.* 3:2.[39] Since the polarisation of emitted light depends on the symmetry of the involved wave functions this underlines the importance of shape control.

An impressive example of shape control has been reported for the growth of ZnO nanocrystals.[40] In the absence of citrate as a shape controlling agent, pencil-like rods were grown, but with increasing citrate concentration the aspect ratio (height to width) of these hexagonal crystals decreased rapidly, suggesting that citrate adsorbs at the basal planes and inhibits growth along the (001) direction (Figure 18).

The shape control using multiple surfactants has not only been expanded to other semiconductor systems such as CdS, ZnS and CdTe but also to certain metals. For example, disc-shaped Co nanocrystals can be produced by inhibiting the growth of the (001) and (00–1) crystal faces using alkyl amines.[39]

Also, gold nanocrystals were obtained with quite perfect tetrahedral, icosahedral and cubic shape by a modified polyol process. The shapes were highly sensitive to the concentration of the gold precursor.[41]

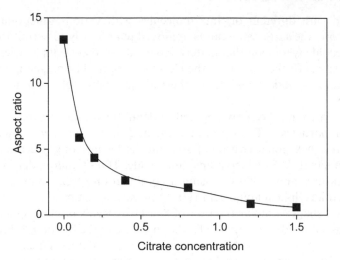

Figure 18 *Aspect ratios (height to width) of hexagonal ZnO nanocrystals as a function of citrate concentration*
(Redrawn based on Ref. 40)

An alternative method of shape control involves the influence of various kinds of inorganic salt additives at different concentrations. From a study of the synthesis of copper nanocrystals, it was proposed that the anion may be the main parameter of shape control, but the crystal structure remained fcc in all cases.[42] It was demonstrated that both chloride and bromide ions adsorb selectively on the (001) and (111) faces, depending on concentration, to favour either crystal growth in the [110] direction or to form cubes. The aspect ratio of copper nanorods reaches a maximum of *ca.* 15 at a concentration of 0.6 mM NaCl or KCl.

Any adsorption lowers the energy of crystal faces and thus the contribution of the surface to the total energy of a crystallite. It would thus appear that growing crystals from the vapour phase should lead to crystallites with a structure that is more likely to be dominated by surface energy and thus more likely to be different from the bulk equilibrium structure. However, crystal growth from solution provides more variables for a subtle control of shape as well as of structure. From experience we know that for example many structural variants of SiO_2 can be grown mostly from solution and often using templates, and the difference in crystal energy of these structures is very small.[43]

ZnO is a unique material that exhibits semiconducting, piezoelectric and pyroelectric properties with potential applications in optoelectronics, sensors and transducers. Using a solid–vapour phase thermal sublimation technique, under controlled temperature and pressure as well as flow rate of the carrier gas, results in a rich variety of single crystal nanostructures, such as belts, springs, spirals, closed rings, rods, combs, bows and cages[44,45] (Figure 19) reminding to some extent of the water system with its variations of snow flake structures.

Remarkably, all these morphologies are based on the wurtzite structure with its hexagonal cell. It consists of alternating layers of positively charged Zn^{2+} cations and negatively charged O^{2-} anions along the hexagonal axis, resulting in a dipole moment and spontaneous polarisation along this axis. The wide range of structures becomes possible by tuning the growth rates along three main growth directions.[45] The main parameter is the local temperature at the point of deposition, but doping additives seem to play a crucial role as well. When the hexagonal axis is the axis of fast growth this results in nanorods (upper right corner of Figure 19). The 'prickling wire' structure (second in the bottom line of Figure 19) has a clear hexagonal axis with 60° angles between the pricks.

There are three different types of nanobelts: one with the hexagonal axis along the fast growing direction; a second one with this axis perpendicular to the plane of the belt, so that the two flat sides of the belt exhibit opposite polarity; and a third one with this axis in plane, but perpendicular to the growth direction.[45] The latter type (upper left panel in Figure 19) has a typical width of 50–300 nm and a thickness of 10–30 nm, while the first type is much narrower, only about 6 nm wide. In order to reduce the electrostatic energy the second type of nanobelt tends to spiral up to form springs with quite an even pitch, and a diameter of 500–800 nm. Nanohelices are much narrower, only about 30 nm in diameter, and made up

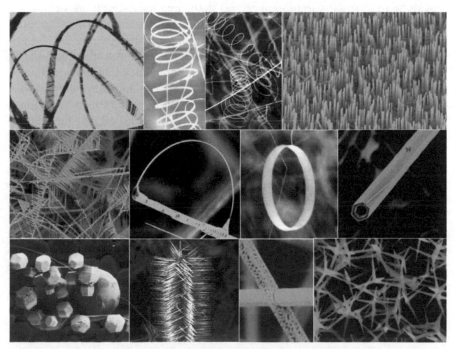

Figure 19 *A collection of mostly single crystal ZnO nanostructures synthesised under controlled conditions by thermal evaporation of solid ZnO powder, partly in the presence of small or equimolar amounts of other oxides*
(Reprinted from Ref. 44. Copyright (2004) with permission from Elsevier. Image provided by Prof. Zhong Lin Wang, Georgia Tech.)

of 12 nm nanowires in a hexagonal screw coiling with a most interesting sequential change of the growth direction that avoids the build up of elastic energy.[46]

When ZnO is heated with a small addition of lithium carbonate and indium oxide, seamless single crystal nanorings with micrometre diameters are observed (third panel in second line of Figure 19).[47] They are formed by self-coiling and subsequent sintering of polar nanobelts (third type) so that the opposite charges of the polar belt surfaces compensate. Reportedly, a key for the fast growth is a planar defect along the growth direction of the nanobelt. It is always present and lowers the nanobelt energy, but does not affect its polarity. Two different types of rings were found, one with the hexagonal axis parallel to the ring axis, the other one with a tilted axis.

Modifying the composition of the source material can drastically change the morphology of the grown oxide nanostructure. When a 1:1 mixture of ZnO and SnO_2 is used, the latter decomposes into Sn and O_2 at high temperatures, and tiny liquid Sn droplets are found at the tips of the pricks or formed nanorods (Figure 20). It is thought that these liquid drops act as a catalyst, serving as a preferential site for absorption of gas phase reactant and, when supersaturated, the nucleation site for crystallisation.[45,48]

Nanobelts are also formed from other oxides including SnO_2, In_2O_3, Ga_2O_3, CdO and PbO_2.[49] SiO_2 forms amorphous nanosprings.

A quite spectacular variation in crystal shape and mesostructural architecture was also reported for electrodeposited lead on highly ordered pyrolytic graphite electrodes as a function of the applied reduction potential (Figure 21).[50] Crystallites of icosahedral, decahedral, octahedral, hexagonal and triangular shapes form at voltages near the equilibrium reduction potential, while higher reduction potentials lead to more complex architectures such as nanowires, nanobrushes, monopods and multipods. The architecture is strongly affected by the growth rates, which can be influenced by the electrolyte concentration and the offset of the potential from its equilibrium value. The polyhedral bodies can be described by a collection of single-crystal tetrahedrons that are twin-related at their adjoining faces. Thus, five tetrahedrons arranged symmetrically about an axis form a decahedron, and 20 tetrahedra in a three-dimensional configuration assembled about a common vertex produce an icosahedron.[50]

Partly, the complex structures that are obtained can be a result of self-assembly, and again surfactants can play an important role in separating crystallites from each other, but at the same time arranging them in chains akin to rows of domino stones, or in two- or three-dimensional superlattices, as has been demonstrated for the case of $BaCrO_4$ nanoparticles.[51] Such regular superlattices are expected to have exciting optical or magnetic properties with applications as photonic crystals or for data storage.

7   Size Effects on Ion Conduction in Solids

Ion conduction is of prime importance for solid state reactions in ionic systems, and for devices such as high temperature batteries and fuel cells, chemical filters

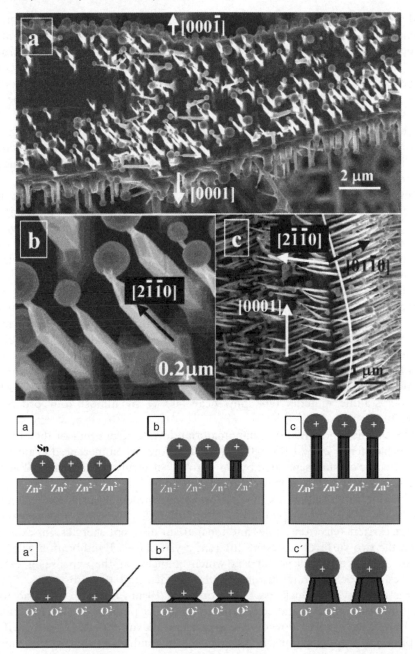

Figure 20 *Crystalline ZnO nanowires and nanorods with liquid Sn drops at the tip (top), and suggested growth mechanism (bottom). Another commonly used catalyst for ZnO is Au*
(Reprinted with permission from Ref. 48. Copyright (2004) American Chemical Society)

and sensors.[52,53] Vacancies facilitate ion transport and thus conductivity in the crystal lattice. In pure materials it is the *configurational entropy* which is responsible for the presence of such vacancies under equilibrium conditions. Due to electroneutrality, each vacancy on the cation sublattice must be compensated by a vacancy of the same charge in the anion sublattice, or by other defects of corresponding charge. The number of such vacancies can be increased by doping the solid with ions of higher charge. In recent years, this principle has provided an important strategy of increasing ion conductivity in the bulk of various materials, and for making them suitable for practical applications. At high temperatures one may obtain very high defect concentrations and a transition to a superionic conducting state.

Inspection of the simplified model shown in Chapter 2, Figure 3 reveals that the energy required to bring a cubic model atom to a kink site (3′) on a surface is higher when the atom is taken from the bulk (where it is stabilised by six coordinating neighbours) as compared to the case when it is taken out of a smooth surface (where it is stabilised only by five neighbours). From this, it follows directly that the statistical probability for the formation of a vacancy is higher for a lattice site on an isolated surface than for one in the bulk.[32] The higher defect concentration in a surface layer suggests that ion conductivity within this surface layer of a low mobility material in general is higher than in the bulk; while the opposite behaviour is found in general for high mobility materials.[32] As for other surface-determined effects the overall (*i.e.* bulk plus grain boundary) conductivity scales with the inverse grain size. The conductivity of nanocrystalline CaF_2 with a grain size of 9 nm exceeds that of the microcrystalline material (200 nm) near 500 K by almost four orders of magnitude.[32]

The chemical potential of a nanomaterial deviates from that of the bulk by the surface term, $2\gamma V_m/r$ (Equation 30). Therefore, when measured against the macroscopic crystals separated by Cu^+ or Ag^+ electrolytes, nanocrystalline Cu and Ag show non-zero electrochemical potential.[32]

A major modification to this picture occurs when a surface of an ionic conductor is brought into equilibrium contact with a second surface where the charge carriers (electrons, holes and ions) are at different energies, for example when the two surfaces represent different crystal faces. Equilibration will lead to an exchange of charge carriers which changes, at the expense of excess charges, the defect concentration in the surface and the subsurface. Space charge zones with a thickness of the order of 10 nm build up and lead to a bending of electronic band edges and to a contact potential which in equilibrium prevents a further flow of charge carriers. We are familiar with this effect in principle from junctions between two semiconductors (where the charge carriers are only electrons and holes but not normally ions).

A special size effect beyond the normal scalable behaviour arises when a layered material is so thin that the space charge zones of opposite surfaces overlap. This was demonstrated by Sata *et al.*,[52] for ionic heterostructures consisting of alternating CaF_2/BaF_2 layers. For these experiments the overall thickness of the stack of layers remained approximately constant near 500 nm,

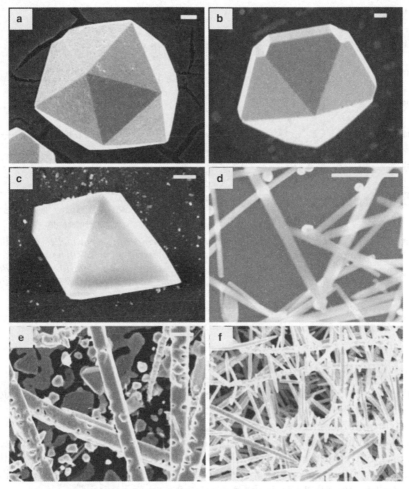

Figure 21 *Structural evolution of Pb nanocrystals with applied reduction potential for solutions containing 5 mM lead nitrate and 0.1 M boric acid. The reduction potentials for (a)–(f) are −0.8, −0.95, −1.1, −1.2, −1.7 and −2.0 V, respectively. The length of the scale bar is 500 nm and the growth time is 60 s (Reprinted with permission from Ref. 50. Copyright (2004) American Chemical Society)*

but the thickness of the individual CaF_2/BaF_2 layers varied from 16 nm to 430 nm. The conductivity parallel to the film is shown in Figure 22. It is obviously significantly larger for the heterostructures than for either bulk BaF_2 or CaF_2 alone. Down to a periodicity of about 100 nm (double layer thickness) we observe a conductivity increase with decreasing thickness that is ascribed to a normal size effect. Below 100 nm down to 16 nm there is a strong increase due to space charge overlap. On further reduction of the periodicity the conductivity drops steeply, which is the expected theoretical behaviour; but it may also be that the films are no longer continuous at these tiny thicknesses.[53]

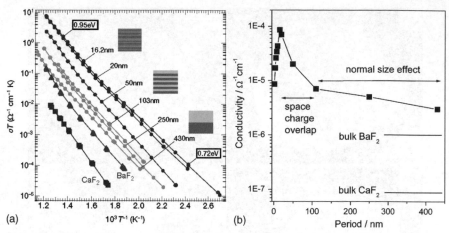

Figure 22   *(a) Arrhenius plot of conductivity times temperature parallel to* CaF_2/BaF_2
                 heterostructure films with a total thickness of 470 ± 40 *nm but varying*
                 periodicity (the numbers denote the thickness of the individual CaF_2/BaF_2
                 double layers). (b) Conductivity of the same heterostructure films as a
                 function of periodicity at 593 K
                 (Reprinted from Ref. 52 with permission from the publisher)

8   Principles of Self-Assembly

Molecular self-assembly is a strategy for nanofabrication that involves design-
ing of molecules and supramolecular entities so that shape-complementarity
and hydrophobic, electrostatic, magnetic or capillary interactions causes them
to aggregate into desired ordered structures. The term has been used in various
contexts, but most commonly it is used for 'processes that involve pre-existing
components (separate or distinct parts of a disordered structure), are reversible,
and can be controlled by proper design of the components'.[54]

Since self-assembly is an equilibrium process from a disordered into a more
ordered structure, we can understand the first basic principle by looking at the
more familiar processes of condensation from the gas phase and crystallisation
from the gas or liquid/solution state, as in

$$NM_{gas} \xrightleftharpoons[\quad]{-\Delta H_V, -\Delta S_V} M_{N,\text{liq.}} \xrightleftharpoons[\quad]{-\Delta H_m, -\Delta S_m} M_{N,\text{cryst.}} \qquad (31)$$

For $N = N_A$ the equilibrium constant K of each step is given by $\Delta G_i^\circ = -RT$
$\ln K_i$, with $\Delta G_i^\circ = \Delta H_i^\circ - T\Delta S_i^\circ$, the suffix ($^\circ$) denoting complete transformation from the unmixed reactant into the unmixed product state. Since conden-
sation and crystallisation are cooperative phenomena there is a first-order
phase transition at a temperature T_i at which $\Delta G_i^\circ = 0$, and thus

$$\Delta H_i^\circ = T\Delta S_i^\circ \qquad (32)$$

Equation (32) demonstrates that the formation of the more ordered state is determined by the interplay between enthalpy and entropy. ΔH_i° and ΔS_i° can be assumed independent of temperature over a not too large range. Since both, entropy and enthalpy decrease from the disordered to the more ordered state we can lower temperature until $T\Delta S_i^\circ$ is lesser than ΔH_i° in magnitude, so that ΔG_i° is negative and the ordered state is thermodynamically stable. It was found empirically that the entropy of evaporation amounts to about 87 kJ mol^{-1} for most liquids at 1 atm. About two-third of this is due to the increase of volume by about three orders of magnitude from liquid to gas. Thus we learn that at a given temperature we can also reach the phase transition to the more ordered state by increasing concentration up to its saturation value.

We may want to have a given system going directly from the gas to the crystalline phase without an intermediate liquid state, a process that we call sublimation. This can be achieved by tuning the boiling point to a value below the melting point, $T_b < T_m$, which is equivalent to requesting

$$\frac{\Delta H_v^\circ}{\Delta S_v^\circ} < \frac{\Delta H_m^\circ}{\Delta S_m^\circ} \tag{33}$$

The parameter that is tuneable to some extent is ΔS_v°. By working at low pressure ΔS_v° can be made larger so that the additional phase may be circumvented. In solution the equivalent of ΔS_v° is the entropy of dissolution of a liquid two phase system. Here we can work at sufficiently low concentration, but it may be easier to change the heat of dissolution by changing the solvent.

The above recommendations often do not provide sufficient tunability for a chosen system. Here is where systems which we tend to call self-assembling, offer additional flexibility. Self-assembly often involves *amphiphiles*, that is, molecules with a charged or otherwise hydrophilic head group and one or more hydrophobic chains. The most common option to tune the relevant thermodynamic parameters involves changing the length or nature of these hydrophobic chains. This permits to draw from extensive experience with intermolecular interactions and structure formation that has been accumulated over several decades in the area of colloids. An excellent introduction has been written by Israelachvili.[14]

We inspect a situation in which monomers in solution form aggregates of different size. Equilibrium thermodynamics requires that for a given solution each monomer has the same chemical potential in all steps of aggregation.

$$\mu = \mu_1^0 + kT \ln x_1 = \mu_2^0 + \frac{kT}{2} \ln \frac{x_2}{2} = \mu_3^0 + \frac{kT}{3} \ln \frac{x_3}{3} = \mu_N^0 + \frac{kT}{N} \ln \left(\frac{x_N}{N} \right)$$

monomers   dimers   trimers   *N*-mers

$$\tag{34}$$

where μ_N is the chemical potential of a monomer in an aggregate with N members, μ_N^0 is the standard chemical potential of the monomer in this aggregate (the mean interaction free energy per molecule), and x_N is the activity based on the mole fraction of molecules in the corresponding aggregate. Equation (34) can be rewritten as

$$x_N = N\left\{x_1 \exp\left[\frac{(\mu_1^0 - \mu_N^0)]}{kT}\right]\right\}^N \tag{35}$$

For $\mu_1^0 = \mu_N^0$ Equation (35) implies $x_N << x_1$, so that most molecules would remain in the monomer state. Thus, the necessary condition for the formation of stable aggregates is $\mu_N^0 < \mu_1^0$.[14] Since at first glance aggregation is entropically unfavourable, this normally means that the molecule needs ent-halpic stabilisation in the aggregate to offset the unfavourable entropic contri-bution. Furthermore, x_N can never exceed unity. Therefore, x_1 cannot exceed further when it approaches a critical value, called the *critical micelle concen-tration* (CMC),

$$x_{1,\text{cmc}} = \exp\left[-\frac{(\mu_1^0 - \mu_N^0)}{kT}\right] \tag{36}$$

The CMC is a monomer solubility limit. Aggregates are formed when the surfactant concentration increases above the CMC. These represent a new phase (or new phases) which remain microdispersed rather than undergoing macroscopic phase separation. For the case of aqueous solutions of surfactant molecules which are transferred into micelles or bilayers, the hydrophobic free energy decreases by an increment of between 1.7 and 2.8 kJ mol^{-1} per CH$_2$ group of the aliphatic chain. The CMC therefore drops by a factor of 2–3 per CH$_2$ group added to the chain.[14] The main contribution to this hydro-phobic free energy increment is of entropic origin, as illustrated by the free energy of transfer of *n*-butane from the bulk disordered liquid into water, which splits up as follows: $\Delta G = \Delta H - T\Delta S = (-4.3 + 28.7)$ kJ mol$^{-1} = +24.5$ kJ mol^{-1}.[14] The origin of the large negative entropy of transfer of a hydro-phobic molecule into water is ascribed to the water structure which breaks up to accommodate the hydrocarbon. Apparently, the reorientation of the water molecules which permits the retention of the network of hydrogen bonds around the solute leads to a more ordered water structure. An alternative explanation ascribes the hydrophobic effect to its characteristic and often extensive compensation of ΔH of solvation and $T\Delta S$ of cavity creation.[55] Negative entropy of solvation is well known also for noble gas atoms and non-polar diatomic molecules.

It should be noted that due to the reduced coordination number of the surfactant molecules at the edge of planar aggregates, for example of bilayers, μ_N^0 becomes size-dependent and decreases progressively with increasing N. As a result, planar aggregates which form above the CMC tend to approach infinite size. This is basically for the same reason that large crystallites in equilibrium with a given solution are thermodynamically more stable than small ones.

The situation is different for monomers, which prefer to aggregate into structures with a curved surface, such as spherical micelles or bilayer vesicles. They may do so for simple geometrical reasons, in particular when their polar head group has a higher cross section than the hydrocarbon tail. Such a geometrical constraint leads to a fixed curvature and thus to a more or less fixed size of the aggregate. It means that μ_N^0 does not decrease monotonically as N increases, but it assumes a minimum at a value N^* which corresponds to the most stable aggregate size. Alternatively, the systems may assume some periodic three-dimensional bicontinuous structure which permits them to retain the optimum surface curvature and still grow to infinite size.

The polar head groups of surfactant molecules are often positively or negatively charged. It is clear that Coulomb forces, and thus the degree of dissociation of the ionic head groups, as well as added electrolyte contributes significantly to μ_N^0 and therefore to CMC values and preferred aggregate structures. Replacement of singly charged by doubly charged counter ions can have a dramatic effect. In addition to the length of the hydrophobic

Figure 23 *Equilibrium structures from molecular dynamics simulations of three-dimensional superlattices at 300 K. Left: body-centred cubic lattice made of [Au_{140} ($C_4H_9S)_{62}$] clusters with the (100) face in front and the upper right corner cut. Right: tetragonally distorted face-centred cubic (bct) lattice made of [$Au_{140}(C_{12}H_{25}S)_{62}$] with (100) face exposed in front. Gold: yellow, sulfur: red, alkyl chains: green. Note the preferentially aligned molecular bundles and their 'interlocking'*
(Reprinted with permission from Ref. 56. Copyright (1996) American Chemical Society)

chain the nature of the head group and the concentration of added electrolyte are parameters which permit the tuning of desired self-assembled aggregate structures. Surfactant molecules have therefore gained tremendous importance in nanotechnology not only because they adsorb selectively at certain
surfaces of crystallites and permit the control of growth rates. Rather, the
various aggregates with their controlled size and often periodicity permit us
to use such phases as templates for the synthesis of mesoporous molecular
sieves in sol–gel processes, and inorganic material dissolved in well-defined
portions in the aqueous part of inverted micelles can be deposited in two-
or three-dimensional periodic arrays and converted to quantum dots of
controlled size.

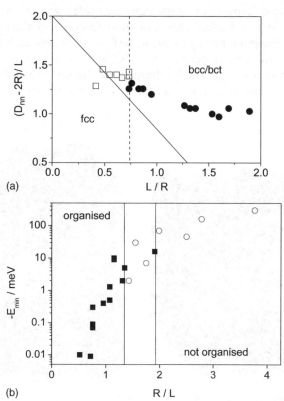

Figure 24 *(a): Experimental structural results plotted against the ratio of alkyl chain
length L to metal core radius R. D_{nn} is the nearest-neighbour distance
(Reproduced from Ref. 57 with permission from the Royal Society of Chemistry) (b): variation of the stabilisation energy against R/L for various
nanocrystal–thiol systems. Ordered structures are represented by closed symbols, non-organised structures by open symbols*
(Reprinted with permission from Ref. 58. Copyright (2000) American
Chemical Society)

Two examples of three-dimensional superlattice assemblies of gold clusters passivated with alkylthiol molecules from molecular dynamics simulations are shown in Figure 23.[56] Although both core–shell structures with their hard core and soft 'wrapping' are near spherical, the two assume different lattice structures. Short alkyl chains lead to body-centred (bcc) structures; moderately long ones obviously induce a tetragonal distortion. The situation reminds of the different packing structures of spherical noble gas or metal atoms (Chapter 3, Section 1). For the present superlattices, the situation was carefully analysed using molecular dynamics simulations and by comparing with experiments,[57] and it was found that the ratio of the alkyl chain length L to the metal core radius R plays a key role. At a ratio of *ca* 0.75, the structure switches between fcc and bcc/bct (Figure 24, upper part).

A similar structural change was observed for two-dimensional arrays of alkylthiol passivated Pd nanocrystals of varying diameters and different alkyl chain lengths (C_4–C_{16}).[58] It was found that for R/L values smaller than 0.75 and greater than 2.0 the nanocrystal arrays are disordered or form low-order structures whereas there is a transition region for ratios between 1.4 and 1.9 (Figure 24, lower part). The corresponding phase diagram was explained by a soft-sphere model that assumes the total interaction potential to be a sum of a steric contribution due to the alkyl chains and an attractive van der Waals contribution due to the polarisation of the metal core (which has a much higher polarisability than the alkyl shell). Short alkyl chain lengths or larger metal cores ($R/L > 1.5$) give steep potential energy wells in the range of tens of milli-electronvolts, possibly implying agglomeration. For $R/L < 0.75$, the two-particle potential exhibits a shallow minimum with negligible stabilisation. The long thiol chains shield the attractive interactions between the metal cores. The structure is dominated by the packing of the chains, leading to low order. For intermediate R/L values in the range ~ 1–1.5, the stabilisation energies have moderate values comparable to the thermal energy of the nanocrystals, which is just adequate to bring the metal particles close enough to form an extended array.[58] A compatible statement by Landman and Luedtke[57] says that the entropic contribution to free energy of the superlattice assembly is found to be large and of similar magnitude as the potential energy component of free energy.

Besides the three- and two-dimensional superstructures discussed above, analogous one-dimensional structures have also been fabricated. Uniform $BaCrO_4$ nanocrystals of 16 nm length and 6 nm width were 'stitched' together by means of surfactant molecules to one-dimensional chains or rectangular two-dimensional arrays.[51] Other work involved micelle-like self-assembled structures using rod-like bipolar metal-polymer amphiphiles.[59] Magnetic superstructures are also of high interest. Face-centred cubic super-lattices of cobalt nanoparticles were reported, and when the structures were grown with an applied magnetic field this led to tubes from crystals coated with citrate ions whereas thick films were obtained when the coating was replaced by dodecanoic acid.[6]

Key Points

- For sufficiently small systems the concept of phases and phase transitions loses its meaning.
- For small isolated systems fluctuations become increasingly important, and the principle that one has to consider only the most probable distribution for a statistical treatment of a system does no longer apply. Furthermore, 'thermal' noise becomes quantised, and temperature is no longer meaningfully defined.
- Melting point and boiling point of free liquid nanosize droplets are below those of the bulk material.
- Liquids which wet a solid surface of a pore are pulled in spontaneously by capillary forces. Inside they have a lower vapour pressure and a lower melting temperature.
- When a sessile drop is in contact with a rough surface or with a structured surface with trapped air in pockets, so that only a small fraction of the exposed area is in contact with the liquid, the surface exhibits superhydrophobicity and repels dirt (Lotus effect).
- The nucleation rate is an extremely strong function of the degree of supersaturation.
- Monodisperse particles can be synthesised by producing a strong supersaturation for a short time and then continue particle growth under conditions of weak supersaturation.
- The higher the initial degree of supersaturation the larger the number of nuclei and the smaller the final particle size (Weimarn's law).
- Thermodynamics dictates a single crystallite morphology described by Wulff's law for a given crystallographic structure of a given material with intrinsic (equivalent to 'as cleaved') surfaces. Large deviations from this principle arise when these surfaces are modified, for example by surface liquid layers or adsorbed inhibitor molecules which reduce the surface free energy and/or suppress the growth rates. Deviations from thermodynamic equilibrium geometries imply growth under kinetic rather than thermodynamic control.
- The formation probability of vacancies is higher in the surface layer of a crystal than in its bulk. In rigid materials this leads to higher particle mobility and higher ion conductivity near the surface.
- When crystal faces of ionic conductors of different chemical potential (for example due to different size) are in contact, it leads to an exchange of charge carriers and to the build-up of space charge zones analogous to those at semiconductor interfaces.
- An aggregated state is more highly ordered than a molecularly dispersed state. By conventional wisdom it should therefore be entropically unfavourable. While this is true in general it is often not true for self-assembly from aqueous solution. Hydrophobic solvation of hydrocarbon chains is associated with *negative* entropy which drives the system to the aggregated state.

General Reading

- T.L. Hill, *Thermodynamics of Small Systems*, Dover, New York, 1994.
- D.E. Gross, *Microcanonical Thermodynamics – Phase Transitions in 'Small' Systems*, Lecture Notes in Physics, vol 66, World Scientific, Singapore, 2001.
- D.J. Evans and D.J. Searles, The fluctuation theorem, *Adv. Phys.*, 2002, **51**, 1529.
- J. Maier, Nano-ionics: trivial and non-trivial size effects on ion conduction in solids, *Z. Phys. Chem.*, 2003, **217**, 415.
- Y. Yin, A.P. Alivisatos, Colloidal nanocrystal synthesis and the organic–inorganic interface, *Nature*, 2005, **437**, 664.
- J. Israelachvili, *Intermolecular and Surface Forces*, Academic Press, London, 1992.
- D. Chandler, Interfaces and the driving force of hydrophobic assembly, *Nature*, 2005, **437**, 640.
- P. Mulvaney, Zeta potential and colloid reaction kinetics, in *Nanoparticles and Nanostructured Films*, J.H. Fendler (ed), VCH, Weinheim, 1998.

References

1. T.L. Hill, *Nano. Lett.*, 2001, **1**, 111, 159, 273.
2. R.B. Shirts, Brigham Young University, personal communication.
3. T.L. Hill, *Thermodynamics of Small Systems*, Dover, New York, 1994.
4. M. Heuberger, M. Zäch and N.D. Spencer, *Science*, 2001, **292**, 905.
5. J. Israelachvili and D. Gourdon, *Science*, 2001, **292**, 867.
6. G.M. Wang, E.M. Sevick, E. Mittag, D.J. Searles and D.J. Evans, *Phys. Rev. Letts.*, 2002, **89**, 050601.
7. S. Kos and P. Littlewood, *Nature*, 2004, **431**, 29.
8. S.A. Crooker, D.G. Rickel, A.V. Balatsky and D.L. Smith, *Nature*, 2004, **431**, 49.
9. M. Hartmann, G. Mahler and O. Hess, *Phys. Rev. Letts.*, 2004, **93**, 080402.
10. R.S. Berry, *Phases and phase changes of small systems*, in *Theory of Atomic and Molecular Clusters*, J. Jellinek (ed), Springer, Berlin, 1999.
11. R.S. Berry, *C.R. Physique*, 2002, **3**, 319.
12. L.D. Gelb, K.E. Gubbins, R. Radhakrishnan and M. Sliwinska-Bartkowiak, *Rep. Prog. Phys.*, 1999, **62**, 1573.
13. G. Zhao, B. Gross, H. Dilger and E. Roduner, *Phys. Chem. Chem. Phys.*, 2002, **4**, 974.
14. J. Israelachvili, *Intermolecular and Surface Forces*, Academic Press, London, 1992.
15. S.J. Gregg and K.S.W. Sing, *Adsorption, Surface Area and Porosity*, Academic Press, London, 1982.
16. W. Thomson, *Phil. Mag.*, 1871, **42**, 448.
17. O. Sinanoglu, *J. Chem. Phys.*, 1981, **75**, 463.

18. Q. Jiang, L.H. Liang and D.S. Zhao, *J. Phys. Chem. B*, 2001, **105**, 6275.
19. J. Weissmüller, *J. Phys. Chem. B*, 2002, **106**, 889.
20. W. Barthlott and C. Neinhuis, *Planta*, 1997, **202**, 1.
21. BASF news. www.corporate.basf.com/basfcorp/corpsfiles/pressefotodb/ 1208_wassertropfen-water_droplets.jpg.
22. R.N. Wenzel, *J. Phys. Colloid Chem.*, 1949, **53**, 1466.
23. A. Otten and S. Herminghaus, *Langmuir*, 2004, **20**, 2405.
24. S. Herminghaus, *Europhys. Lett.*, 2000, **52**, 165.
25. J. Kim and C.-J. Kim, *Technical Digest, IEEEConference on MEMS*, Las Vegas, 2002, 479.
26. X. Feng, L. Feng, M. Jin, J. Zhai, L. Jiang and D. Zhu, *J. Amer. Chem. Soc.*, 2004, **126**, 62.
27. T. Sun, G. Wang, L. Feng, B. Liu, Y. Ma, L. Jiang and D. Zhu, *Angewandte Chemie. Int. Ed.*, 2004, **43**, 261.
28. A.W. Adamson, *Physical Chemistry of Surfaces*, Wiley, Toronto, 1982.
29. V.J. Anderson and H.N.W. Lekkerkerker, *Nature*, 2002, **416**, 811.
30. S. Eiden, *Nachrichten aus der Chemie*, 2004, **52**, 1035.
31. G. Wulff, Z. Krist. Kristallgeom., 1901, 949.
32. J. Maier, *Z. Phys. Chem.*, 2003, **217**, 415.
33. C.T. Campbell, S.C. Parker and D.E. Starr, *Science*, 2002, **298**, 811.
34. H. Reiss, *J. Chem. Phys.*, 1951, **19**, 482.
35. Y. Yin and A.P. Alivisatos, *Nature*, 2005, **437**, 664.
36. X. Peng, J. Wickham and A.P. Alivisatos, *J. Amer. Chem. Soc.*, 1998, **120**, 5343.
37. Y. Furukawa, *Chemie in unserer Zeit*, 1997, **31**, 58.
38. S. Mann, *Angew. Chemie*, 2000, **112**, 3533.
39. E.C. Scher, L. Manna and A.P. Alivisatos, *Phil. Trans. R. Soc. London A*, 2003, **361**, 241.
40. Z.R. Tian, J.A. Voigt, J. Liu, B. McKenzie, M.J. McDermott, M.A. Rodriguez, H. Konishi and H. Xu, *Nature Materials*, 2003, **2**, 821.
41. F. Kim, S. Connor, H. Song, T. Kuykendall and P. Yang, *Angew. Chem.*, 2004, **116**, 3759.
42. A. Filankembo, S. Giorgio, I. Lisiecki and M.P. Pileni, *J. Phys. Chem. B*, 2003, **107**, 7492.
43. P.M. Piccione, C. Laberty, S. Yang, M.A. Camblor, A. Navrotsky and M.E. Davis, *J. Phys. Chem. B*, 2000, **104**, 10001.
44. Z.L. Wang, *Materialstoday*, June 2004, p. 26.
45. Z.L. Wang, *J. Phys.: Condens. Matter*, 2004, **16**, R829.
46. R. Yang, Y. Ding and Z.L. Wang, *Nano. Lett.*, 2004, **4**, 1309.
47. X.Y. Kong, Y. Ding, R. Yang and Z.L. Wang, *Science*, 2004, **303**, 1348.
48. P.X. Gao and Z.L. Wang, *J. Phys.Chem B*, 2004, **108**, 7534.
49. Z.L. Wang, *Ann. Rev. Phys. Chem.*, 2004, **55**, 159.
50. Z.-L. Xiao, C.Y. Han, W.-K. Kwok, H.-H. Wang, U. Welp, J. Wang and G.W. Crabtree, *J. Amer. Chem. Soc.*, 2004, **126**, 2316.
51. M. Li, H. Schnablegger and S. Mann, *Nature*, 1999, **402**, 393.
52. N. Sata, K. Eberman, K. Eberl and J. Maier, *Nature*, 2000, **408**, 946.

53. J. Jamnik and J. Maier, *Phys. Chem. Chem. Phys.*, 2003, **5**, 5215.
54. G.M. Whitesides and B. Grzybowski, *Science*, 2002, **295**, 2418.
55. G. Graziano, *Phys. Chem. Chem. Phys.*, 1999, **1**, 3567.
56. W.D. Luedke and U. Landman, *J. Phys. Chem.*, 1996, **100**, 13323.
57. U. Landman and W.D. Luedtke, *Faraday Discuss.*, 2004, **125**, 1.
58. P.J. Thomas, G.U. Kulkarni and C.N.R. Rao, *J. Phys. Chem. B*, 2000, **104**, 8138.
59. S. Park, J.-H. Lim, S.-W. Chung and C.A. Mirkin, *Science*, 2004, **303**, 348.
60. M.P. Pileni, Y. Lalatonne, D. Ingert, I. Lisiecki and A. Coutry, *Faraday Discuss.*, 2004, **125**, 251.

Adsorption, Phase Behaviour and Dynamics of Surface Layers and in Pores

1 Surface Adsorption and Pore Condensation

1.1 The Langmuir Adsorption Isotherm

Adsorption isotherms are most important for the characterisation of the surface area and the pore structure of large surface area materials. They are derived at first for flat surfaces, but it is for curved surfaces, in particular for nanopores, where they obtain their significance for size-dependent phenomena.

An adsorption isotherm gives the adsorbed amount n of a species, normally in units of a monolayer coverage n_m, as a function of the relative pressure p in units of the vapour pressure p_0 of the bulk liquid at the same temperature.

The simplest example is the Langmuir isotherm (also called type I isotherm). It treats strong adsorption, in particular chemisorption. Its main characteristic is that the maximum surface coverage is that of a monolayer. There are normally specific adsorption sites, and each site can bind a single molecule. This situation is represented by the equation

$$\frac{n}{n_m} = \frac{cx}{1 + cx} \tag{1}$$

where $x = p/p_0$ is the relative pressure of the adsorbate in the gas phase and c is a constant that characterises the adsorption equilibrium. For $cx >> 1$, the curve saturates at $n = n_m$ (Figure 1).

1.2 The Brunauer–Emmett–Teller (BET) Equation

The basic model for weakly adsorbing (physisorbing) atoms and molecules was developed by Brunauer, Emmett and Teller.[1] A monolayer is built up, and a second and third layer follow. Depending on the relative magnitude of the adsorption energy on the substrate and the adsorbate condensation energy, the second layer starts to build up before the first one is complete (see Figure 2). The first layer often has a preferential orientation which is coverage dependent and dictated by the competition between the specific interaction of the adsorbate with

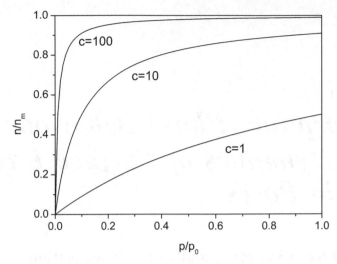

Figure 1 *The Langmuir isotherm of strong adsorption (chemisorption) where the max-imum coverage corresponds to a monolayer*

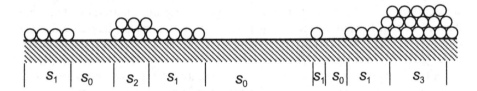

Figure 2 *The BET model of adsorption, shown for a nominal monolayer of the adsorbate*

the substrate and the interaction among the adsorbate molecules. Under most circumstances the memory of the substrate surface is lost after only a few layers.

The basic assumption is that the Langmuir equation applies to each layer. The heat of adsorption for the first layer is Q_1. All further layers adsorb on top of the adsorbate, the heat of adsorption is therefore taken to be equal to the heat of condensation or vaporisation of the adsorbate, Q_v. The top layer is assumed to be in dynamic equilibrium with the layer underneath. The rate of condensation on an exposed surface area s_i with i layers is then equal to the rate of evaporation from an area s_{i+1}, with corresponding rate coefficients a_1 and b_i, respectively. Thus, for condensation on s_0 and evaporation from s_1 we have

$$a_1 p s_0 = b_1 s_1 \exp\left(-\frac{Q_1}{RT}\right) \tag{2}$$

and for all succeeding surfaces

$$a_i p s_{i-1} = b_i s_i \exp\left(-\frac{Q_v}{RT}\right) \tag{3}$$

Following Adamson[2] and abbreviating

$$y = \frac{a_1}{b_1} p \exp\left(\frac{Q_1}{RT}\right) \tag{4}$$

and substituting

$$x = \frac{a_i}{b_i} p \exp\left(\frac{Q_v}{RT}\right) \tag{5}$$

we can write

$$s_1 = ys_0, \qquad s_2 = xs_1 \tag{6}$$

and

$$s_i = x^{i-1} s_1 = yx^{i-1} s_0 = cx^i s_0 \tag{7}$$

where

$$c = \frac{y}{x} = \frac{a_1 b_i}{b_1 a_i} e^{(Q_1 - Q_v)/RT} \cong e^{(Q_1 - Q_v)/RT} \tag{8}$$

Summing up the adsorbed amounts and dividing by the monolayer coverage (total available sites on the surface) we obtain

$$\frac{n}{n_m} = \frac{\sum\limits_{i=1}^{\infty} i s_i}{\sum\limits_{i=0}^{\infty} s_i} = cs_0 \frac{\sum\limits_{i=1}^{\infty} i x^i}{s_0 + s_0 c \sum\limits_{i=1}^{\infty} x^i}. \tag{9}$$

Insertion of the algebraic equivalents for the sums yields

$$\frac{n}{n_m} = \frac{cx/(1-x)^2}{1 + cx(1-x)} \tag{10}$$

which rearranges to the familiar expression

$$\frac{n}{n_m} = \frac{cx}{(1-x)[1 + (c-1)x]} \tag{11}$$

Taking the ratio of rate coefficients (a_i/b_i) to be the same as for the liquid adsorbate – vapour equilibrium identifies x as equal to p/p_0.

Figure 3 shows the behaviour of Equation (11) for various values of c. The initial slope at small values of x is a measure of c. The curve for $c = 100$ with its knee near $n = n_m$ is typical for nitrogen or argon on the surface of many solids for which the heat of adsorption is significantly larger than the heat of condensation Q_v. It is called a type II isotherm. The one for $c = 0.1$ is of type III and is obtained for $Q_v > Q_1$.

The BET equation has become the standard for determining surface areas, usually with nitrogen at 77 K as the adsorbate with a molecular cross section of

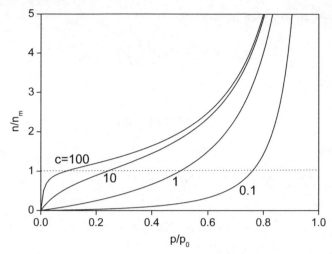

Figure 3 *BET isotherms as a function of the relative vapour pressure for different values of c which is a measure of the wetting power of the adsorbate. The dotted line at $n/n_m = 1$ represents a nominal (average) monolayer*

0.16 nm^2. It is often also written in a rearranged form as

$$\frac{x}{n(1-x)} = \frac{1}{cn_{\mathrm{m}}} + \frac{(c-1)x}{cn_{\mathrm{m}}} \tag{12}$$

which gives a linear plot of the left-hand side versus x, so that n_{m} and c can be obtained from the slope and intercept. For non-porous materials the linear range that indicates validity of the BET assumptions extends typically between $0.05 \leq x \leq 0.3$, but for microporous materials it may be limited to $x \leq 0.1$.

For $c \cong 100$ or more, an approximate form of Equation (11) is often used.

$$\frac{n}{n_{\mathrm{m}}} = \frac{1}{1-x} \tag{13}$$

This permits an estimate of the surface area based on n_{m} from a single point, which is normally chosen near $x = 0.25$ for non-porous powders and near $x = 0.1$ for microporous materials because deviations from BET behaviour are often observed at higher adsorbate pressures.

1.3 Adsorption in Micropores

Adsorption in wide pores is the same as on a single flat surface, but as soon as the opposite surface of a slit pore comes into reach of the van der Waals potential, typically at a critical distance of *ca.* 0.5 nm, it leads to additional stabilisation of the adsorbate. This is demonstrated in Figure 4 for an adsorbate in a slit pore, with Lennard-Jones potential parameters (Chapter 3, Equation 2) which represent approximately the interaction of N_2 with a graphite surface. At a distance of the Lennard-Jones parameter σ ($\sigma = 0.4$ nm in the present

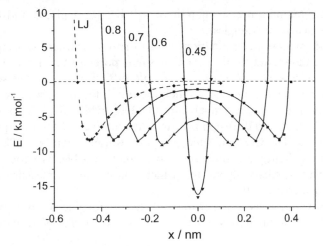

Figure 4 *Interaction potential profile for Lennard-Jones (12-6) particle with 0.4 nm diameter and $\varepsilon_0 = 8.3$ kJ mol^{-1} in a slit-shaped nanopore of the indicated diameter between 0.45 nm and 0.8 nm. LJ gives the same potential on a single surface (shifted to the left for clarity). The curves are spline interpolations between the calculated points*

example) attractive and repulsive terms balance. For a 12-6 potential the minimum, given by ε_0, is at a distance of $2^{1/6}\sigma = 1.122\sigma$ near a single flat surface. For the double-sided surface of a slit pore we obtain a double minimum which decreases in energy the narrower the pore is, until at a pore width of $1.122\sigma = 0.45$ nm, the minima from the two Lennard-Jones branches collapse to a single minimum of double depth. When the pore width narrows further the minimum moves up rapidly because the repulsive terms take over.

There are two major points to which we need to pay attention. First, if we have a cylindrical pore instead of a slit pore this increases stabilisation because the forces operate around the entire circumference of the spherical adsorbate. Second, these adsorption energies are not sufficient for an accurate prediction of adsorbate equilibria. In particular, for low mass adsorbates we have to take into account their zero-point vibrational energies. They increase in narrower pores. Furthermore, there is a significant entropic contribution which is particularly important for light particles in narrow pores. It increases with temperature and works against a localisation of the particle inside the pore. Nevertheless, adsorption into narrower pores begins at a much lower pressure than on a flat but otherwise identical surface.

For micropores with a size below ≈ 2 nm we should speak about pore filling rather than condensation. Ultramicropores with a width of only a few molecular diameters fill at far lower pressures due to the increased interaction energy. Due to the spatial constraints adsorbate molecules have to assume coordination numbers below those in the bulk liquid. The driving force for adsorption will therefore be dramatically different. It may even be that the critical point above which there is no distinction between gas and liquid drops below the experimental temperature so that there will be no meniscus and the Kelvin equation is clearly meaningless.

The experimental determination of adsorption isotherms using nitrogen at 77.4 K is hampered by the fact that small pores of 0.5–1 nm diameter fill at pressures as low as $p/p_0 \approx 10^{-7}$–10^{-5}, and diffusion is very slow so that long equilibration times are needed. Under these conditions it is of advantage to use argon at liquid argon temperature (87.3 K) which fills the same pores at $p/p_0 \approx 10^{-5}$–10^{-3}, and diffusion is faster. Alternatively, CO_2 adsorption at 273 K may be used. The CO_2 molecule is slightly thinner than N_2 (0.28 vs. 0.3 nm radius) so that slightly narrower pores become accessible. Moreover, diffusion is much faster at this temperature, which again reduces measuring times by more than a factor of 10.

Quantitative analysis of adsorption isotherms for microporous materials is often based on the phenomenological *Dubinin–Radushkevich* (DR) equation

$$\log_{10}(n) = \log_{10}(n_p) - D \log_{10}^2(p^0/p) \tag{14}$$

where n is the adsorbed amount, n_p is the pore capacity and D is an empirical parameter. A plot of $\log_{10}(n)$ against $\log_{10}^2(p^0/p)$ should give a straight line which extrapolated to $p = p^0$ gives the micropore capacity n_p.[3]

A novel method for determining the micropore size distribution was introduced by Horvath and Kawazoe (HK). It is based on the idea that the necessary pressure at which a micropore fills is directly related to the potential energy between adsorbate and adsorbent. It was originally derived for slit-like graphitic pores[4] but has since been extended to argon and nitrogen adsorption in various zeolite and aluminophosphate pores.[3] It assumes that the enthalpic term dominates adsorption in small pores. It takes into account the specific adsorbate–adsorbent interaction parameters and pore shapes (see e.g. Figure 4) and derives a direct analytic relation between relative pressure and pore size.

A more recent strategy involves computer simulation of the isotherms based on Grand Canonical Monte Carlo (GCMC) or density functional (DFT) calculations. They can take care of adsorbate–adsorbate interaction and in principle take into account the complex shape and force field of the adsorbate as well, although calculations for particles of spherical symmetry are of course much easier. Nevertheless, they still have to assume that the pores are all of the same simple shape, but given this is true they can determine pore size distributions which should be considerably more reliable than those from phenomenological equations.

1.4 Adsorption and Condensation in Mesopores

Often, the initial width of a pore is larger than the critical width discussed above, so that adsorption at first follows BET behaviour, which allows the determination of a meaningful surface area. When the layers build up in the pores the free space shrinks, and at some point the interaction with the opposite surface increases stabilisation to an extent that the surface layers become unstable and collapse to form a plug (see below). This raises an adsorption isotherm to a level above its BET value, so that Equation (11) no longer provides an adequate description in this range of the isotherm.

Various methods have been developed since 1945 and are still being used to cope with the situation of pore condensation. Derived for specific rigid and well-defined pore geometry, they are either cylindrical, slit-shaped, spherical or the interstices between packed spheres, and the derived pore size distribution depends on this choice. They are based on the Kelvin equation which is assumed to be valid over the entire mesopore range (2–50 nm). The Kelvin radius at which pore condensation of nitrogen ($V_m(l) = 34.71$ cm^3 mol^{-1}, $\gamma_{lv} = 8.85$ mN m^{-1}) at 77.35 K occurs is related to the pressure as

$$r_K = \frac{-0.415 \, \text{nm}}{\log_{10}(p/p_0)} \tag{15}$$

which means that nitrogen condenses in a cylindrical pore with a radius $r_0 = 2$ nm at $p/p_0 \approx 0.6$. Just before condensation the multilayer thickness under these conditions amounts to $t \approx 0.6$ nm, which is a significant fraction of r_0, showing that the corrections for t are essential. The 5 nm pores will fill at correspondingly higher relative pressures, $p/p_0 \approx 0.8$, and $t \approx 0.9$ nm.

The most popular method of pore size analysis for micropores is the BJH method by Barrett, Joyner and Halenda,[5] even though it underestimates pore sizes below ≈ 7.5 nm seriously. The deviations may be as much as 20–30%. This may be due to the limited validity of the Kelvin equation. In these narrow pores the meniscus curvature may no longer be determined solely by the pore size and shape. Effects of the wall which is often rugged on a molecular scale may influence curvature in the thinner part of the meniscus, and the surface tension may be different from its bulk value.[3]

A probably more reliable approach which circumvents these classical macroscopic models is based on DFT or Monte Carlo (MC) calculations. It automatically takes account of the density modulations near the pore walls. Neither surface tension nor contact angle has to be assumed. It looks as if these procedures would become the methods of choice in future. Simulated adsorption isotherms of N$_2$ on model porous glass with known pore size distribution were analysed on the basis of the BJH method. Comparison with the exact distribution revealed that the BJH analysis gave overly sharp distributions that were systematically shifted to lower pore sizes by about 1 nm (20–30%).[6]

1.5 Determination of Mesopore Volumes and Mean Pore Size

Adsorption isotherms of mesoporous materials display a distinct plateau when all pores are filled. The integrated amount of adsorbate up to this plateau is a measure of pore capacity. Under the assumption that the liquid has the same packing density in the pores as it has in the bulk the adsorbed amount can be converted into a total pore volume. Furthermore, given that the surface area is already known from the low pressure part of the isotherm, a mean radius of cyclindrical pores is readily calculated by dividing double the specific pore volume by the specific BET surface area. In the case of slit pores with parallel walls the same procedure yields the pore width.[3] These parameters are useful for a rough but easy characterisation of a porous system. For the case of micropores, since the packing density of the adsorbate is different, pore volumes and pore sizes derived this way have to be regarded with care.

2   Adsorption Hysteresis and Pore Criticality

Much can be learned about architecture, size and surface of pores from *adsorption isotherms*. Argon and nitrogen are used most often as probe molecules, and experiments are conducted at fixed temperature, commonly at the boiling temperature of the probe gas at standard pressure p_0, but other temperatures are used as well. Isotherms are plotted as the amount of the adsorbed gas in various units against the equilibrium pressure, normally in units of p_0.

Let us consider the different stages of condensation of a vapour in a long cylindrical pore (pore radius r_p) that is open on both sides:

- We start with a clean surface and zero vapour pressure, and we assume that the adsorption energy of the molecule on the initial pore surface is greater than the bulk condensation energy of the adsorbate molecule. On increasing the vapour pressure we will initially have molecular adsorption as described by Brunauer, Emmet and Teller (BET adsorption).
- Further adsorption occurs still following BET on top of a liquid (or solid) of the same composition as the vapour, therefore the contact angle θ vanishes (*i.e.* cos $\theta = 1$). The statistical thickness t of the layer increases slowly, and the residual pore radius is $r_p - t$.
- When the external pressure reaches a value that corresponds to the pressure given by the Kelvin equation at the Kelvin radius $r_1 = r_K = r_p - t$ ($r_2 = \infty$ in Equation 7 of Chapter 6), condensation sets in and the pore fills instantaneously. Ideally, for monodisperse pores, we obtain a vertical section in the adsorption isotherm (in principle the isotherm should bend back, but this is not physically realised). In a non-monodisperse pore size distribution the small pores fill first and the pore size distribution is obtained from the slope of the isotherm.
- Alternatively, since the film that builds up is metastable and becomes increasingly unstable it may collapse and block the pore *before* r_K is reached. Thereby, a fraction of the material in the pore is collected under formation of a meniscus with $r_1 = r_2$, as shown in Chapter 6, Figure 2b. From now on condensation occurs with a proceeding meniscus, since the vapour pressure above the doubly curved surface is lower, and condensation on the meniscus is favoured over that of the cylindrical film.
- Desorption from filled pores occurs from a receding hemispherical meniscus ($r_1 = r_2$ in Equation 7 of Chapter 6) and thus over some fraction of the isotherm at a lower vapour pressure than adsorption for which $r_2 = \infty$. This leads to *hysteresis loops* in the isotherms, which are typical for porous materials. It is generally accepted that the shape of the hysteresis loop relates

to the pore size distribution and architecture of the mesoporous adsorbent, but the details of the interpretation are still under discussion. It is clear that thermodynamically, hysteresis is always related to a non-equilibrium effect. In this picture the thermodynamic equilibrium regime is the *desorption* branch of the hysteresis loop, while *adsorption* occurs in the metastable film state. The desorption branch is often near vertical, much steeper than the adsorption branch. This holds in particular for *ink bottle* type pores with large inside diameter but narrow exit window. This is because the liquid desorbs from the narrow exit at a lower pressure than the inside, so that the pore empties instantaneously as soon as the window is free. For cylindrical pores the desorption branch is more parallel to the absorption branch, but it does not have the maximum offset of a factor of $e^2 = 7.4$ that might be expected on the basis of adsorption onto a cylindrical film ($r_2 = \infty$) and desorption from the hemispherical meniscus ($r_2 = r_1$).

A fascinating *in situ* dynamic study of condensation, evaporation and transport of water inside the nanopores of carbon nanotubes (CNTs) using environmental electron scanning microscopy was reported by Rossi *et al.*[7] The CNTs were placed inside an evacuated chamber with controlled temperature and water vapour pressure. When the pressure increased above 5 Torr at 4°C menisci began to become evident inside the tubes, with contact angles of *ca.* 5–20°, before the first indication of water condensation on the sample holder. This shows that the CNTs were more hydrophilic than steel (which has a contact angle of *ca.* 55°). It is likely that the graphene sheets in the walls are hydrogen terminated as a result of synthesis in a C–H environment. Surface tension after the formation of a thin film of water inside the tube applies a force that is sufficient to contract the tube slightly.[7] It is furthermore believed that the forces responsible for liquid being drawn inside a hydrophilic CNT are surface tension and *thermocapillarity*. The latter represents a type of *Marangoni effect*, which describes mass flow due to a gradient in the surface tension. In thermocapillarity this gradient is induced by a temperature gradient. In the more familiar example of tears in a glass of alcohol it results from ethanol evaporating where the liquid wets the sides of the glass so that its surface tension increases and pulls up a liquid film and contracts it to form tears.

Figure 5 reports the change in meniscus shape in response to water vapour pressure for the example of a water plug of approximately 10^{-17} L. The menisci are asymmetric, probably because the tube cross section is not perfectly circular. Furthermore, the shape of the meniscus at the right, the side where the nanotube is closed, is more complex because the slightly different pressure of the trapped bubble exerts a force on the plug. Furthermore, the amount of liquid seems to decrease as the pressure increases. This was ascribed to the low scan rate which leads to a lag in the imaging of the dynamic processes.

It is instructive to verify the picture based on real but simple model adsorption isotherms such as the ones displayed in Figure 6. Adsorption of Ar at 87 K in controlled pore glass (CPG) with a pore diameter of 15.7 nm is shown in the

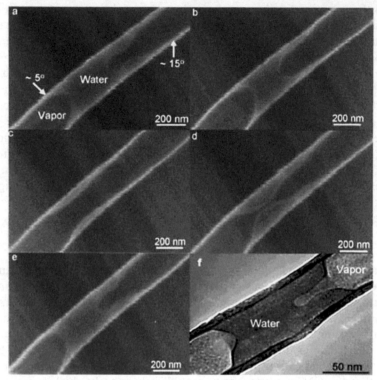

Figure 5   *Environmental scanning electron microscopy images of a water plug in a carbon*
nanotube. The meniscus shape changes when, at constant temperature, the
vapour pressure of water in the experimental chamber is changed from 5.5 Torr
(a) to 5.8 Torr (b), 6.0 Torr (c), 5.8 Torr (d) and 5.7 Torr (e) where the
meniscus returns to the shape seen in (a). TEM image of a similar plug shape in
a closed carbon nanotube under pressure is shown in (f).
(Reprinted with permission from Ref. 7. Copyright (2004) American Chemical Society)

upper entry. The isotherm starts with a flat linear increase of the adsorbed
amount. This is the behaviour of a gas with no significant adsorption (for
comparison: the amount of an ideal gas in a glass bulb at constant temperature
is also a linear function of pressure). At a pressure of 0.9 p_0 there is essentially a
vertical increase which indicates a first-order phase transition; in the present
case it indicates pore condensation. If we had Ar in a glass bulb at 87 K (the
boiling point of Ar) then this increase would occur at $p/p_0 = 1$, indicating a gas–
liquid transition, and the increase would stop when the bulb is filled. In the
same way, from the plateau near $p = p_0$ we can read directly that 1 g of this
sample of CPG has a pore volume of about 0.5 cm^3.

The isotherm of the MCM-48 silica sample with a pore diameter of 2.67 nm
looks much different. The steep unresolved initial increase demonstrates that Ar
adsorbs strongly on the surface. This increase is often followed by a relatively flat

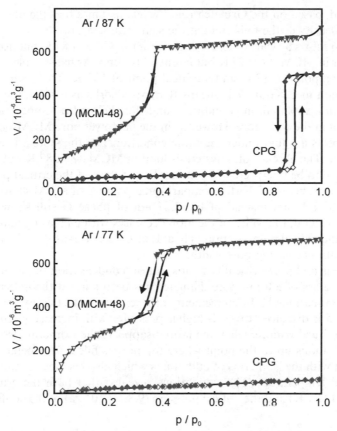

Figure 6 *Adsorption–desorption behaviour for Ar at 77 K in controlled pore glass (CPG) of 15.7 nm pore diameter, and in a phase D of the MCM-48 microporous material (2.67 nm pores).*
(Reprinted with permission from Ref 9. Copyright (2000) American Chemical Society)

section in the pressure range $0.05 < p/p_0 < 0.3$ which corresponds roughly to an adsorbed monolayer from which the surface area is determined on the basis of the known atomic or molecular cross section of the adsorbate and the BET theory. In the present case this results in a surface area of 1060 m^2 g^{-1}, but the flat section is not well expressed because of the onset of pore condensation, and the BET theory is therefore inaccurate. It is followed by a steeper but not vertical increase at a pressure just below $0.4\,p_0$. This step occurs at a pressure much lower than for the CPG material, which is interpreted on the basis of the Kelvin equation (Equation (6.7)) in terms of condensation in pores of much smaller diameter. The fact that the step is not vertical is normally analysed on the basis of a pore size distribution, but it should also be reminded that phase transitions are not expected to be sharp in nanoscopic systems, even for a monodisperse size distribution. From the final plateau value a pore volume of 0.88 cm^3 g^{-1} is determined, based on the Gruvitch rule,[8] which assumes that the pore fluid has the same density as the bulk fluid at the same temperature.

Figure 6 shows a clear hysteresis loop for Ar in CPG at 87 K but not at 77 K, and for Ar in MCM-48 at 77 K but not at 87 K. Bulk Ar has a triple point of 84 K, a boiling point of 87 K and a critical point of 157 K. The absence of pore condensation in CPG at 77 K means that Ar adsorbs as a solid, therefore the system is in a sublimation equilibrium and thus below the temperature of the triple point in the wide pore. However, in the narrower pore MCM-48 material Ar adsorbs as a liquid under the same conditions, so the system is above the triple point. The absence of a hysteresis loop in MCM-48 at 87 K but not at 77 K may have to be interpreted by a dramatic decrease of the critical point from 150.7 K to just below 87 K (for comparison: a phase C of MCM-48 with a pore diameter of 2.81 nm instead of the 2.67 nm of phase D still shows a small loop[9]). In general, the width of the loop decreases with increasing temperature, *i.e.* when the critical point is approached. It also decreases as the critical point is lowered with decreasing pore width.

Hysteresis and pore criticality in adsorption/condensation of Ar in mesoporous silica glass of 3.0 nm pore diameter has been studied theoretically using GCMC simulations.[10] In agreement with experiment, the hysteresis loop shrinks and is displaced towards higher pressures with increasing temperature (Figure 7). Furthermore, when the loop disappears the isotherm also becomes more continuous up to the point where the pore is filled. This behaviour is in agreement with the isotherms of bulk fluids which also become continuous near the critical temperature. The simulations revealed that both the critical temperature (132.7 K) and the critical pressure (9.85×10^5 Pa) of the confined fluid

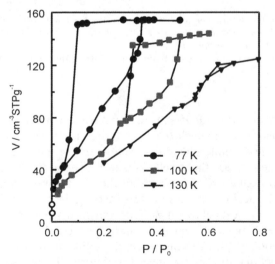

Figure 7   *Hysteretic coexistence curve resulting from GCMC simulations of bulk and of confined fluid argon in mesoporous silica glass of 3.0 nm diameter.*
(Data selected from Ref. 10. Reproduced with permission from the PCCP Owner Societies)

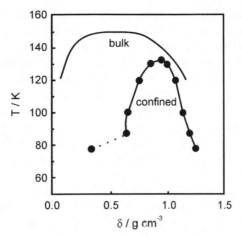

Figure 8 *Hysteretic coexistence curve for the bulk and the confined fluid argon.*
(Data selected from Ref. 10. Reproduced by permission of the PCCP Owner
Societies)

were below the corresponding bulk values of 150.7 K and 1.75×10^6 Pa,
respectively. The gas branch of the hysteretic coexistence curve of the confined
fluid, and therefore also the critical point, is shifted to the higher density region
because it contains both the gas molecules in the core of the pores and the
adsorbed molecules (Figure 8). Other fluids confined in ordered and disordered
porous solids show a very similar behaviour.[10]

Important in the context of hysteresis is the pore size-dependence of phase
transition temperatures: for a wetting liquid the boiling point in a pore is
increased, while melting point, triple point and critical point decrease. The
reason for an often dramatic decrease of the critical temperature may be sought
in the spatial constraint which does not permit the maximum possible coordination number and therefore reduces the maximum total interaction energy per
molecule in a condensed state.

The lowering of the critical point below its bulk value is predicted from mean
field theory. The critical temperature is given by

$$T_c = c \frac{z\varepsilon}{k} \qquad (16)$$

where z is the magic coordination number that is responsible for many size-
dependent phenomena, ε the interaction energy with a nearest neighbour molecule and c is a constant.[11] For a confined fluid the average coordination number
will be lower than for the bulk, depending on the extent of the spatial constraint.
Besides various proposals to describe the effect, Fisher and Nakanishi[12] used
scaling arguments to show that for large pores the decrease of the critical
temperature should be

$$\Delta T_c = \frac{T_c - T_{pc}}{T_c} \propto R^{-1/\nu} \qquad (17)$$

where v is the critical exponent for the correlation length which adopts a value of 0.63 for bulk three-dimensional fluids and 1 for small pores.

The pore critical temperature, T_{pc}, is defined as the temperature where the sharp jump in adsorption isotherms due to capillary condensation just disappears. It is sometimes implicitly assumed that it is the same temperature where hysteresis disappears, but there is evidence that these two temperatures are different, with $T_{pc} > T_h$.[12] This is plausible, since hysteresis results from the system being trapped in a local minimum of the free energy. Thus it does not reflect equilibrium conditions. Morishige and Shikimi[13] estimated $T_{pc} \approx 98$ K and $T_h \approx 62$ K for Ar in siliceous MCM-41 of 1.2 nm pore radius. As expected for small pores, both these temperatures are substantially below the bulk fluid critical temperature of Ar, which is 150.7 K (Figure 9).

In an attempt to explain the hysteresis phenomenon, Cohen[15] proposed that for cylindrical pores hysteresis should occur only when the pores are open on both sides but not when they are closed on one side since in the latter case the meniscus could build up already at the beginning of adsorption. This assumption was tested in a recent experiment of nitrogen adsorption on well-defined non-interconnected pores. Comparison was made between samples with pores which were open on one side with virtually identical ones except that the pores were open on both sides.[16] Hysteresis was found in both cases, so that Cohen's proposal was clearly disproved.

Phase transitions of matter confined in pores can be discussed on the basis of results of experiments and of computer simulations which in a simple case lead to a schematic *p–T* diagram as given in Figure 10. With decreasing pore radius the triple point shifts progressively to lower temperatures and pressures.

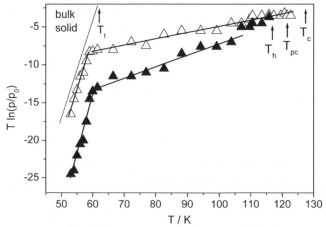

Figure 9   *Difference of chemical potential for capillary condensation (open symbols) and evaporation (closed symbols) of N_2 in controlled pore glass (11.5 nm) with respect to the bulk liquid. T_c: bulk critical temperature, T_{pc}: pore critical temperature, T_h: disappearance of hysteresis and T_t: bulk triple point.*
(Data selected from Ref. 14 and reprinted with permission. Copyright (2004) American Chemical Society)

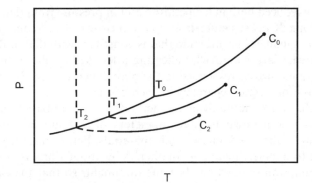

Figure 10   *Schematic p–T-diagram of a bulk fluid with triple point T_0 and critical point C_0, and of a fluid confined to pores of decreasing size.*
(Reprinted with permission from Ref. 9. Copyright (2000) American Chemical Society)

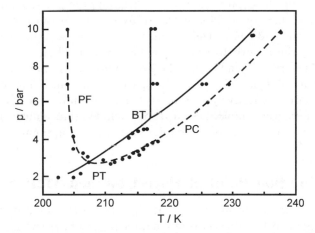

Figure 11   *Phase diagram of CO_2 in Vycor glass. The solid curve refers to the bulk and the dashed curve to the pore-confined system (PF: pore freezing, PC: pore condensation). BT and PT are the triple points for bulk- and pore-confined systems, respectively.*
(Reprinted with permission from Ref. 17)

Simultaneously, the critical point above which no condensation to a liquid is possible is also reduced both in temperature and pressure.[9] At a given pressure, the melting point decreases in the pore whereas the boiling point increases with decreasing pore radius. More detailed studies reveal a more complex behaviour involving new phases and phase transitions (see Chapter 7, Section 4.2).

An experimental example where the complete phase diagram was mapped out has been reported by Duffy *et al.*[17] for CO_2 in Vycor glass with a mean diameter of 7 nm (see Figure 11). The confinement shifts the liquid–solid transition to a considerably lower and the gas–liquid transition to a higher temperature. The triple point is reduced both in temperature and in pressure from that of the bulk. In a microscopic X-ray structure determination Brown *et al.*[18] concluded that the confined CO_2 solidifies in a structure that is consistent with that of the bulk, but with a somewhat larger unit cell, indicating a less dense solid in the pore. The average crystallite size corresponded to the pore dimensions of Vycor, indicating that crystallisation occurs separately in each of the pores.

It seems that bulk thermodynamic methods which are based on the Kelvin equation work well down to pore diameters of 7–8 nm in porous carbon materials and to about 3.5 nm for oxide materials. Below this, the macroscopic models tend to underestimate pore diameters by about 1 nm. In narrow pores, attractive fluid–wall interactions become dominant, so that the classical concept of a smooth liquid–vapour interface and a bulk-like core fluid is no longer realistic.[9] In this regime, microscopic (molecular) descriptions such as DFT, MC and molecular dynamics (MD) simulations are more successful and provide an accurate description. A quantitative explanation of hysteresis, however, is still a matter of debate.

In context with the depression of the critical point in pore-confined material we should also mention a calorimetric study of benzene adsorbed on zeolite NaY at several loadings.[19] The absence of any phase transitions and the magnitude of the adsorbate heat capacities were attributed to a state that resembles supercritical conditions. The low coordination number which leads to the lowering of T_c is enforced by the spatial constraint in the zeolite pores. The zeolite lattice effectively acts as a spacer to enforce a dispersion and clustering of benzene molecules at room temperature similar to the one encountered in the absence of the zeolite near the critical point.

3   The Melting Point of Pore-Confined Matter

For the decrease of the melting point T_m of matter confined in pores an equation similar to Equation (15) of Chapter (6) can be derived:[20]

$$\frac{T_m - T_m^*}{T_m^*} = \frac{\Delta T_m}{T_m^*} = -\frac{fV_m(l)(\gamma_{ws} - \gamma_{wl})}{\Delta H_m r} \qquad (18)$$

Here, f is a form factor derived from the Young–Laplace equation ($f = 1$ for long cylinder pores with end effects neglected, $f = 2$ for spherical clusters, and for the end cap hemispheres in a cylindrical pore). γ_{ws} and γ_{wl} are the wall-solid and wall-liquid interfacial tensions, and ΔH_m is the molar melting enthalpy, again often assumed to be the same as the bulk value. A similar expression that also takes account of the difference between the densities of the solid and liquid phases had been derived previously for free clusters by Pawlow.[21] According to the above equation, the melting point in the pore will be lowered when the wall prefers the liquid phase, $\gamma_{ws} - \gamma_{wl} > 0$, and increased when it is the other way around (see Figures 12 and 13). In the limit of complete wetting (zero contact angle of the liquid), $\gamma_{ws} - \gamma_{wl} = \gamma_{sl}$, the wall is no longer involved at the surface of the solid so that the situation becomes equivalent to that of a free cluster

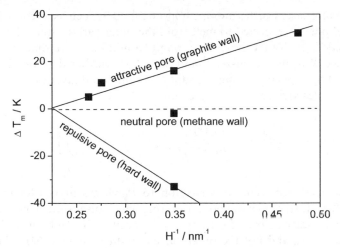

Figure 12 *Shift in freezing temperature from simulation of Lennard-Jones methane in a pore of width H with attractive (graphite), neutral (methane) and repulsive (hard plates) wall.*
(Based on data in Ref. 22)

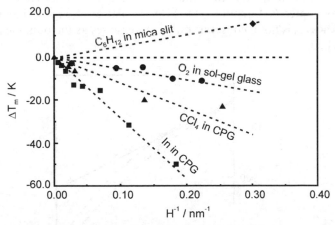

Figure 13 *Shift in experimental freezing point of pore-confined matter as a function of the inverse pore size.*
(Reproduced with permission from Ref. 11)

which is covered by a liquid layer, and the expression reduces to the long-known *Gibbs–Thomson* equation (Equation 15 of Chapter 6).

It should be noted that Equation (18) applies only for the onset of melting. After a short initiation period there will be a liquid layer between the pore solid and the pore wall, and $\gamma_{ws} - \gamma_{wl}$ has to be replaced by γ_{sl} for most of the melting process in any case. Furthermore, a hollow cylinder, such as a frozen layer in a cylindrical pore, is expected to melt from the outer surface since the curvature at the inside is such that the melting point should be above that of the bulk.

Reiss and Wilson,[23] Hanszen[24] and Sambles[25] further developed the liquid shell model by explicitly taking account of a thin liquid layer of fixed thickness t_0, and Sambles arrived at the expression

$$\frac{\Delta T_m}{T_m^*} = -\frac{2V_m(s)}{\Delta H_m}\left\{\frac{\gamma_{sl}}{r-t_0} - \frac{\gamma_{lv}}{r}\left(1 - \frac{\rho_s}{\rho_l}\right)\right\} \tag{19}$$

The last term represents a correction to Pawlow's first-order theory and accounts for a further melting point lowering which disappears when liquid and solid have the same density. Furthermore, it specifies that only the radius of the solid core, $r - t_0$, and not the outer surface radius r enters. MD simulations seem to confirm this expression (Figure 14).

For $\rho_s \approx \rho_l$ and therefore $V_m(s) \approx V_m(l)$, Equation (19) simplifies to

$$\frac{T_m - T_m^*}{T_m^*} = \frac{\Delta T_m}{T_m^*} = -\frac{2V_m(l)}{\Delta H_m}\frac{\gamma_{sl}}{(r-t_0)} = -\frac{K}{r-t_0} \tag{20}$$

which differs only in the modified radius from the traditional Gibbs–Thomson equation (Chapter 6, Equation 15). MD simulations suggest that the liquid layer thickness t_0 is not a constant; rather it increases as the bulk melting point is approached.

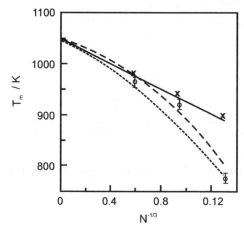

Figure 14 *Computed melting point of gold nanoclusters for a chosen potential as a function of $N^{-1/3}$. The solid line follows Pawlow's theory with $t_0 = 0$, the crosses include second-order corrections for $t_0 = 0$, dashed curves are for Samble's liquid shell theory with $t_0 = 0.5$ nm in the absence (heavy dashed curve) and with second-order corrections included (dotted curve).*
(Reprinted with permission from Ref. 26. Copyright (2001) American Chemical Society)

Equation (19) is often used to estimate the solid–liquid interfacial tension, γ_{sl}, a quantity which is very difficult to be determined directly from the size-dependence of the melting temperature.[26] Empirical rules include the rule by Antonov[27] which derives from Young's equation (Chapter 6, Equation 5) in the limit of complete wetting ($\theta = 0$)

$$\gamma_{sl} = \gamma_{sv} - \gamma_{lv}, \tag{21}$$

and the one by Tyson and Miller[28]

$$\gamma_{sl} = \alpha \cdot \gamma_{sv} \quad \text{and} \quad \gamma_{sl} = (1 - \alpha) \cdot \gamma_{lv} \tag{22}$$

where $\alpha = 0.15 \pm 0.03$ for metallic systems.

The traditional way of investigating the melting behaviour is by differential scanning calorimetry (DSC). Figure 15 shows typical DSC traces based on the example of indium confined in Vycor and in controlled pore glasses of different pore diameter.[29] Each trace shows two peaks. There is a strong and relatively sharp signal with a position that is independent of the pore size. In the limit of

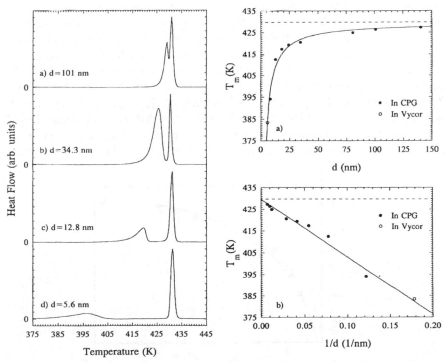

Figure 15 *Left: Background-subtracted DSC melting curves for indium in controlled-pore-glass (a–c) and in Vycor samples (d) with different pore diameters. Each scan was carried out at a heating rate of 10 K min⁻¹. Right: Melting temperature as a function of pore diameter and of inverse pore diameter. The broken line represents the bulk melting point.*
(Reprinted with permission from Ref. 29. Copyright (1993) by the American Physical Society)

zero heating rate it is found to occur at a temperature of 429.6 K, in good agreement with that of bulk indium (429.8 K). This signal corresponds to indium outside the pores, surrounding the CPG grains. The second signal shows two typical size-dependent effects: the peak position shifts to lower and lower temperature the smaller the pore diameter, and at the same time the peak width increases, demonstrating that the melting transition gets less and less sharp. Since the pore size is well-defined the broadening of the feature for the smaller pores cannot be ascribed to a size distribution. Rather, it reflects the fact that melting is a *cooperative phenomenon*, and a sharp melting point is only obtained in the thermodynamic limit of an infinite number of particles.

ΔH_m was determined separately from the calorimetric measurements and found to amount to only 36 and 52% of its bulk value for pores of 4.1 and 9.1 nm, respectively. This permits an independent determination of the interfacial tension, which is also dramatically reduced from its bulk value.[29]

A detailed study of melting and freezing of water in the uniform cylindrical pores of ordered mesoporous silica materials (MCM-41 and SBA-15) was reported by Schreiber, Ketelsen and Findenegg.[20] The melting point depression as a function of pore size is displayed in Figure 16. A single-parameter fit based on the traditional Gibbs–Thomson equation cannot reproduce the experimental trend below 2.5 nm radius, but inclusion of a liquid layer with thickness t_0 as a second parameter leads to a perfect fit with $t_0 = 0.38$ nm. Parameter K in Equation (19) adopts a value of 0.190(7) nm, which agrees within error with its theoretical value that is obtained using bulk properties ($\gamma_{sl} = 32$ mJ m^{-2}, $\Delta H_m = 6.0$ kJ mol^{-1}, $T_m^* = 273.2$ K and $V_m(l) = 18 \times 10^{-6}$ m^3 mol^{-1}). This agreement supports the notion that ΔH_m and γ_{sl} adopt their bulk values, and that there is a liquid shell of 1–2 layers of water molecules which does not contribute to ΔH_m. The conclusion is at the first glance at variance with the above earlier study of indium in Vycor and

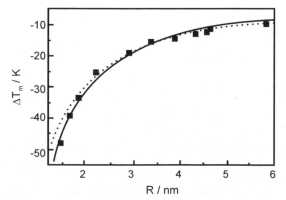

Figure 16   *Shift of the melting temperature for water in fully loaded MCM-41 and SBA-15 materials as a function of pore radius. The full curve is a fit to the data of the modified Gibbs–Thomson equation with $t_0 = 0.38$ nm, the broken curve is a fit assuming $t_0 = 0$.*
(Data selected from Ref. 20. Reproduced by permission of the PCCP Owner Societies)

controlled pore glass, but the two studies are not exactly comparable: the indium experiment relates ΔH_m to the entire amount of metal in the pore whereas the water study excludes a liquid shell so that ΔH_m corresponds only to the melting of the remaining solid core. Furthermore, it is only the ratio between interfacial tension and melting enthalpy which is obtained from the Gibbs–Thomson equation. If both properties are reduced by a comparable amount, as was found in the indium study, then one needs a reliable separate determination of at least one of the two properties. It is in particular a possible size-dependence of the interfacial tension which is of considerable interest.

The question of the size-dependence of the interfacial tension is accessible from the size-dependence of the lattice parameter (see Chapter 2, Equation 9). It was addressed by Goldstein *et al.*[30] for free CdS nanocrystals. For $\gamma(111)$ they obtained a value of 2.50 N m^{-1} which is considerably larger than the bulk value of 0.750 N m^{-1}. This is consistent with similar observations for Al,[31] Pt and Au,[32] where the surface tension of the nanocrystals was also considerably enhanced over the corresponding bulk values. A small increase can be explained by the small contraction of the surface that is reflected by the reduction of the lattice constant. On a contracted surface there are more atoms per unit area which can form a bond to an adsorbate. For metals, the situation can be more complicated because of the metal-to-nonmetal transition which reduces the extent of delocalisation of conduction electrons and makes them more easily available for the formation of bonds to an adsorbate. In this context it is somewhat puzzling that the study by Unruh *et al.*[29] gave evidence for a decrease rather than an increase of surface tension for nanosize indium clusters. As an overall conclusion one may state that the extent of the size-dependence of interfacial tensions is not yet quantitatively understood.

Melting always results in a single calorimetric peak related to pore-confined water. For completely filled pores, the same is true for the freezing transition, but the transition is shifted to lower temperature by 5–10°, with some fluctuation, indicating a hysteresis between melting and freezing which is to be distinguished from supercooling (see below). In contrast, for partly filled pores, freezing in the same materials often exhibits several peaks, depending on the degree of loading (see Figure 17 for an SBA-15 sample with a pore diameter of 7.7 nm). A first peak (denoted by I) is narrow, has a sharp onset at its high-temperature side and occurs only in the sample with a loading corresponding to $\phi = 1.3$ times the pore capacity. It relates to the freezing at 258 K of the supercooled excess water between the crystallites. A second peak (II) in the same sample is observed at 248 K. It is broader and nearly symmetric and is ascribed to the freezing of pore water. A related peak is also observed for incompletely filled pores above some minimum loading, such as for $\phi = 0.6$ in Figure 17b. In this sample the peak related to freezing of pore water is shifted to somewhat lower temperature (245 K), but the sharp onset again indicates supercooling. Interestingly, there are two further peaks at 236 and 233 K (designated III and IV). These two peaks are also observed at $\phi = 0.40, 0.25$ and 0.20, while there is only a weak indication of peak II at $\phi = 0.40$ in its unshifted position. Thus, there is a minimum pore filling near $\phi = 0.40$ below which peak II does not occur. It was found that the area of peak III relative to that

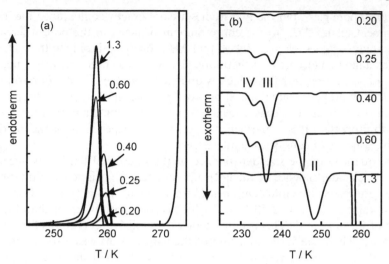

Figure 17 *Differential scanning calorimetric curves for water in SBA-15 with pores of 7.7 nm diameter at different degrees of pore filling ϕ, in units of the full pore capacity: (a) endotherms show the melting peak that depends only weakly on the loading; (b) exotherms exhibit several freezing peaks, depending on the degree of pore filling.*
(Data selected from Ref. 20. Reproduced by permission of the PCCP Owner Society)

of IV decreased with decreasing loading. Furthermore, the freezing temperature of these two peaks is only weakly dependent on pore size or on the degree of loading.

The interpretation of the freezing peaks is based on the loading dependence and rests on a simple model sketched in Figure 18: peak II relates to the freezing of filled pores. The presence of excess water induces freezing so that super-cooling is only found for $\phi < 1$. Peak III was attributed to freezing of a liquid film of adsorbed water at the walls of incompletely filled pores. A thin film is a two-dimensional system for which the lowering of the phase transition temperature is expected to depend primarily on its thickness rather than on its curvature (see Chapter 9, Section 2). The interpretation of peak IV is unclear. An assignment to freezing of the contact layer with thickness t_0 is tempting, but this would be inconsistent with the interpretation of the melting temperatures which seemed consistent with the assumption that this layer does not freeze. Another question that has to be answered concerns the number of peaks: why are there two or more peaks on freezing while there is only a single well-defined feature on melting? It may be that the solidified film evaporates, transports via the vapour phase, and deposits as a solid that fills the pore and shows the same melting behaviour as peak II. Little is known at present about the rate of such a mass transport in a confined system.[20]

A simple and relatively rapid method to determine pore size distributions uses nuclear magnetic resonance and is therefore called *NMR-cryoporometry*.[33] It is based on the fact that the spin–spin relaxation time (also called transverse or

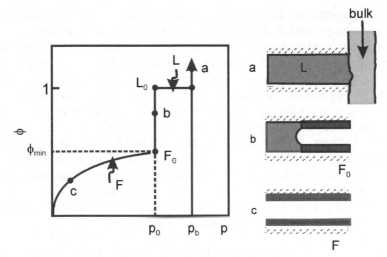

Figure 18 *Sketch of an idealised adsorption isotherm expressing the degree of filling φ as a function of vapour pressure p (left), and model of pore freezing corresponding to points a, b, and c on the isotherm (right). In the low pressure section of the isotherm a fluid film F forms at the pore walls. At a critical pressure p_0 and a multilayer film F_0 pore condensation sets in, leading to a vertical rise of the isotherm and a coexistence of pore liquid L and F_0.*
(Data selected from Ref. 20. Reproduced by permission of the PCCP Owner Society)

T_2 relaxation time) of a solid is short (typically of the order of 1 ms for the protons of cyclohexane near its melting point) but long for liquids (about 1 s for cyclohexane). The free induction decay signal of a pulsed proton NMR experiment of pore-confined cyclohexane can clearly be seen to consist of two distinct components at all temperatures below the normal melting point. The component with the long-time constant is due to the liquid and the short-lived component due to the solid fraction. After correction for the Curie law that characterises the nuclear magnetisation and for the temperature-dependence of the spectrometer response (the quality factor of the receiver coil), the relative amplitudes of the two components directly give the solid and liquid volume fractions of the sample. The main advantages over calorimetric methods are that measurements can be made on well-equilibrated samples and a pore size distribution can be determined. The latter was found to coincide well with those obtained from the analysis of gas desorption experiments.

4   Layering Transitions

4.1   Layering of Solids and Liquids Adsorbed on Smooth Surfaces

At relatively low temperatures adsorption of gases on smooth surfaces can give rise to step-wise increase of the adsorbed amounts in adsorption isotherms. This is generally interpreted as layer-wise adsorption. Apparently, the adsorbing gas prefers to complete each successive monolayer before the next layer is started. We are familiar with a similar behaviour in crystal growth, where it is explained well on the basis of energetic considerations since it is more favourable for an incoming particle to attach at a kink than to start a new layer. However, the layering to be discussed here has to be distinguished from this normal surface effect where at a given temperature the surface is in equilibrium with a single vapour pressure, whereas different number of adsorbate layers correspond to different equilibrium vapour pressures. These adsorbate layers, therefore, have to be regarded as different phases. There are often peculiarities when adsorption occurs not too far below the melting point of the bulk adsorbate so that phenomena like surface roughening and surface wetting transitions occur. Normally, the adsorbate is different from the substrate, and this requires that additional aspects such as lattice mismatch have to be considered.

Methane adsorption on MgO (100) surfaces shows steps in the isotherm at 87.4 K up to the completion of the third layer, after which the isotherms become smooth. When experiments are carried out at a lower temperature of 77 K the transition to smooth behaviour is found after the fifth layer. Quasi-elastic neutron scattering experiments show that these strongly adsorbed first layers are solid-like (the bulk melting point of methane is 90.8 K). Oxygen adsorbed on graphite exhibits similar layering transitions at low temperature with the number of observed transitions depending on temperature. These layers are crystalline below 43.8 K and liquid-like at higher temperatures, and various phase transitions were characterised both below and above the triple point temperature.[12]

Argon adsorption on graphite was studied extensively using X-ray and neutron scattering, adsorption isotherm measurements and ellipsometry.[12] The results of ellipsometric determinations of adsorption isotherms are displayed in Figure 19. While the macroscopic Frenkel–Halsey–Hill model predicts a monotonic increase of the layer thickness, the experimental isotherm shows a

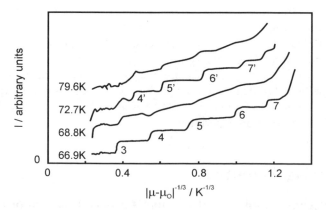

Figure 19   *Ellipsometric coverage isotherms for solid argon on graphite at four temperatures. The abscissa is a function of the reduced pressure, $\mu_n - \mu_0 = T \ln(p_n/p_0)$. (Reproduced with permission from Ref. 12)*

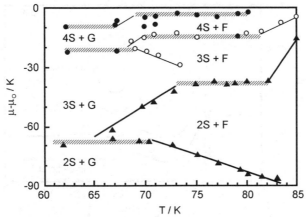

Figure 20   *Phase diagram for argon multilayers on graphite. The hatched lines indicate first-order phase transitions (vertical isotherm steps). S, F, and G abbreviate solid, fluid and gas phase, respectively. Layer critical temperatures are located at $T_C(2) = 70.0$ K, $T_C(3) = 68.3$ K, $T_C(4) = 67.2$ K, and the layer triple points at $T_T(2) = 66.5$ K, $T_T(3) = 67.2$ K. In this diagram, adsorption isotherms from Figure 19 are represented by vertical lines at the given temperature (not shown). It is obvious that for the isotherm near 69 K only the first two layers are well defined.*
(Reproduced with permission from Ref. 34. Copyright (1993) American Chemical Society)

'quantised' behaviour with a regular staircase of seven steps at 66.9 K. At 68.8 K, 15 K below the bulk melting point of Ar, the steps damp out much more quickly, and already after the second layer the isotherm is monotonic. Surprisingly, there is a second series of layering transitions at higher temperature, with steps which are offset in pressure from the first set. This behaviour indicates that a completed liquid-like layer can be solidified by additional adsorption. An estimation of the complete adsorption phase diagram is given in Figure 20.

4.2   Layering Transitions of Confined Fluids in Smooth Pores

In a similar way as on one-sided solid surfaces layering is also expected to be induced by smooth walls of pores. Such effects have not yet been accessed by experiment, but molecular simulations have provided detailed insight.

The results of GCMC simulations of adsorbed Lennard-Jones methane in an attractive graphite slit pore is shown for two temperatures in Figure 21. The simulations reveal three phases, with first-order phase transitions. In all three phases there is a clear layering. For phase A, the in-plane pair correlation functions of all layers damp out rapidly, which indicates the structure of a liquid (at 130 K, upper part in Figure 21). For phase B, shown at 123 K (lower part), the contact layer at the two pore walls shows a clear solid-like structure, and phase C (below 114 K, see Ref. 35) corresponds to all layers being frozen. The freezing point of bulk methane in the same simulations is 101 K. The

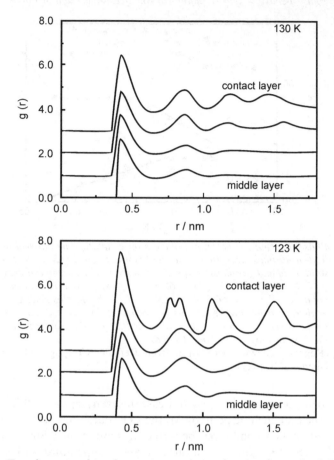

Figure 21 *Two-dimensional in-plane pair correlation function for each of the adsorbed methane layers in the graphite slit pore with a width of H = 7.5σ. Phase A at 130 K (upper) and phase B at 123 K (lower).*
(Reproduced from Ref. 36 with permission from Taylor & Francis Ltd (www.tandf.co.uk/journals))

calculations thus predict that the system represents a less common case where the melting temperature *increases* under confined conditions. The origin of this behaviour is the attractive potential (see also Figure 2 of Chapter 3).

Another comprehensive study of an associating fluid in a slit pore using the DFT method has been reported by Huerta *et al.*[37] The interaction potential consists of an isotropic Lennard-Jones-12-6 potential, characterised by the particle diameter σ and the attractive energy ε. Superimposed on this is an anisotropic potential that simulates the associative character (*e.g.* hydrogen bonding) along tetrahedral directions with a strong square-well attractive potential $\varepsilon_{as} = 30\varepsilon$. There is also a strong Lennard-Jones- 9-3 potential with $\varepsilon_{gs} = 24\varepsilon$ representing the interaction of the fluid with the wall. This ensures that the adsorbate wets the adsorbing surface.

A selection of adsorption isotherms for this system is given in Figure 22. Each of the isotherms exhibits more than one discontinuity. Proceeding along increasing or decreasing overall density branches allows to enter the metastability regions and to evaluate the hysteresis loops. Equilibrium phase transitions are localised by calculating and analysing the excess grand canonical thermodynamic potential (Figure 22d). The different phases are investigated by plotting the density profiles across the pore (Figure 23). Curve (3) in Figure 23a relates to the point marked with label (3) in Figure 22. It obviously represents a filled pore, and except for the first density peak near the pore wall the density profile is nearly uniform. Profile (2) at intermediate overall density shows a high

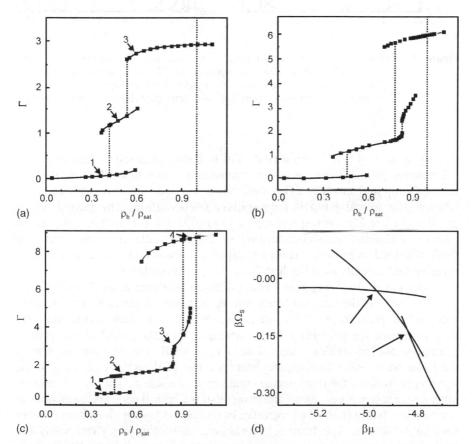

Figure 22 *Adsorption isotherms for the four-site associating Lennard-Jones fluid in slit-like pores at a reduced temperature $T^* = 4kT/\varepsilon$. The pore width D is 5, 10 and 15 in units of the Lennard-Jones parameter σ in parts a, b and c, respectively. Vertical dashed lines correspond to surface phase transition points and the line most to the right corresponds to the bulk gas–liquid coexistence. Part d illustrates the localisation of the phase transition points by using the excess grand potential versus the chemical potential for the pore width $D = 5\sigma$.*
(Reprinted with permission from Ref. 37. Copyright (2000), American Institute of Physics)

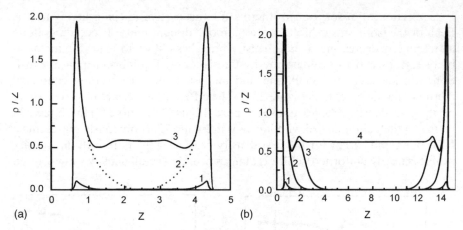

Figure 23 *Density profiles of a Lennard-Jones fluid across slit pores of width D = 5σ (a) and 15σ (b) at T* = 4kT/ε. The lines represent the phases for which the adsorption isotherms are shown and denoted with the same labels in Figs 22a and 22c.*
(Reprinted with permission from Ref. 37. Copyright (2000), American Institute of Physics)

density in the first layer, with a peak that is almost identical to that of profile (3), whereas profile (1) represents an almost empty first layer. This means that the first jump in the isotherm (Figure 22a) corresponds to a layering transition, whereas the second jump is for capillary condensation. The second set of profiles (Figure 23b corresponding to Figure 22c) for $D = 15\sigma$ exhibits an additional layering transition, leading to a second adlayer. Non-associating fluids adsorbed on attractive substrates appear to show much more pronounced layering that extends over the first two, three or more layers.[37]

A selection of phase diagrams for four different slit dimensions, $D = 15, 12.5, 5$ and 4.5 in units of the Lennard-Jones parameter σ, is displayed in Figure 24, along with the bulk phase diagram. Two successive slit widths are shown in the same plot to demonstrate the influence of this parameter. For the model of the present calculation the bulk critical temperature is $T_c^* = 4.6$ K. The diagrams relating to the wider pores reveal three distinct branches, two of them corresponding to the layering transitions, the third one to capillary condensation. The critical temperature of the layering transitions are lower than the critical capillary condensation temperature, but the critical temperature of the second layering transition is lower than for the first one. The different branches meet at triple points. Remarkably, the triple point temperature between the second layering transition and the capillary condensation transition *increases* with decreasing pore width.

With decreasing pore dimension the second layering transition becomes metastable with respect to capillary condensation. The triple point increases further until the entire phase branch connected with this transition disappears. For $D = 5\sigma$ and 4.5σ there are only two branches, and the critical temperature for the layering transition is now equal to or even higher than the critical temperature for capillary condensation. This means that at sufficiently high

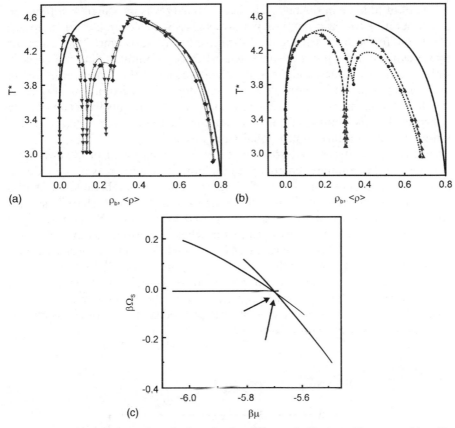

Figure 24 *Phase diagrams for a fluid confined in different slit-like pore. The pore width is 15σ 12.5σ (a) and 5σ and 4.5σ (b). In both cases the triangles correspond to the wider, the other symbols to the narrower pore, whereas the solid lines correspond to the bulk phase diagram. (c) Illustrates the localisation of the triple point from the behaviour of the excess grand thermodynamic potential for a pore width of 4.5σ.* (Reprinted with permission from Ref. 37. Copyright (2000), American Institute of Physics)

temperature, or equivalently at weaker attraction, only layering transitions but no capillary condensation can be observed in sufficiently narrow pores, even if the association between the fluid species is high.[37]

MC simulations of water in cylindrical pores were reported by Brovchenko *et al.*.[38] A decrease of the critical temperature by 35% compared to the bulk value was found in pores with a radius of 1.2 nm.

5   Liquid Coexistence and Ionic Solutions in Pores

Mixtures of 2,6-lutidine (2,6-dimethyl-pyridine) and water display a miscibility gap with a lower critical temperature of 306.8 K at a critical weight fraction

$x = 0.29$ of lutidine in the bulk. The heat capacity was measured as a function of temperature for these mixtures in porous glass with a characteristic pore size of 100 nm for both critical and off-critical compositions (Figure 25).[39] Even for these relatively large pores the onset temperatures of immiscibility were clearly shifted to higher values, and the heat capacity peaks at the phase transition are much less sharp. This demonstrates that size-dependent phenomena also occur for these less common types of phase transitions. The results were consistently interpreted in the context of finite-size scaling theory and supported by calculations based on a three-dimensional Ising model.[40]

Electrolyte solutions are also expected to exhibit different properties under confinement in nanopores compared to the bulk. The main reason is a reduced coordination number relative to that in the bulk, either because of spatial constraints or because of adsorption of the ions at the wall surface. The local structure and coordination is best investigated by extended X-ray absorption fine structure (EXAFS). It was found for RbBr nanosolutions in carbon-slit pores that the Rb–OH_2 distance and the coordination number decreased in a similar fashion for both the 0.7 and 1.1 nm pore widths.[41] In contrast, the hydration structure around the confined Br ions was different in the two pores, with a somewhat lower coordination number in the larger pore. The water molecules removed from the hydration shell formed stable and ordered clusters. The situation is depicted in Figure 26.

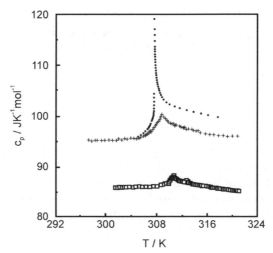

Figure 25   *Phase separation of pore-confined liquid mixture with miscibility gap. Specific heat of 2,6-lutidine (2,6-dimethylpyridine) + water in controlled pore glass of 100 nm average pore size at weight fractions $x = 0.16$ (squares) and $x = 0.29$ (crosses) of lutidine as a function of temperature. The upper (dotted) curve corresponds to a bulk mixture at $x = 0.29$.*
(Reproduced from Ref. 39. Copyright (1993) with permission from Springer)

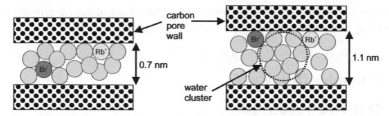

Figure 26 *Hydration model for RbBr nanosolutions obtained from EXAFS measurements for activated carbon with slit pores of two different widths.*
(Reproduced by permission from Ref. 41. Copyright CSIRO 2003, Published by CSIRO PUBLISHING, Melbourne, Australia)

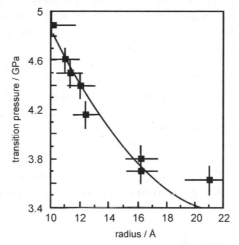

Figure 27 *Size-dependence of the wurtzite to rock salt pressure-induced structural transformation of CdSe at room temperature.*
(Reprinted with permission from Ref. 42. Copyright (1996) American Chemical Society)

6 The Effect of Pressure

Also pressure-induced phase transitions are size-dependent. This has been demonstrated for the solid–solid transformation between the normal wurtzite to the high pressure rock salt structure of CdSe (as shown in Figure 27) and for other compounds. In all cases, as the nanocrystal size decreases, the pressure required to induce the transition to the higher density phase at a given temperature increases, with a scaling law similar to the one that applies for melting, but opposite in direction.[42]

As described by the Laplace–Young equation, $\Delta P = 2\gamma/R$ (Chapter 2, Equation 6), high energy surfaces exert a considerable pressure on a particle with small radius R. It was shown that diamond and not graphite is the thermodynamically stable phase for small clusters of 10^4–10^5 carbon atoms. This permits to grow diamond films as a metastable phase by means of low pressure chemical vapour deposition.[43] Also the cubic high pressure modification of boron nitride is more stable than its hexagonal allotrope below $R \approx 1.6$ nm.[44]

7   Dynamics in Pores

7.1   Dielectric Properties

Dynamic properties of a solvent in confined geometry can be very different from that in the bulk. Experimental studies of water confined in reverse micelles and biological pores have revealed a substantial decrease of polarity and a considerable slowing down of the rate of relaxation of water molecules. Also MD simulations have indicated that water molecules in such environments exhibit a greater degree of spatial and orientational order than in the bulk, and the formation of molecular layers near the cavity surface.[45]

The dielectric constant of water confined in a spherical nanocavity of radius R_c with non-interacting walls was calculated by Senapathi and Chandra[45] based on MD simulations of the mean square moment fluctuation $<M^2>_R$ in a concentric sphere of radius $R \leq R_c$. M is the collective dipole moment of all water molecules within the sphere of radius R. Interactions of all dipolar molecules are included. $<M^2>_R$ therefore includes the effect of the reaction field in the shell $R < r < R_c$, and as a result it depends on the dielectric constant and also on the ratio R/R_c. Berendsen showed that $<M^2>_R$ is related to the dielectric constant ε of the solvent in the cavity by the expression

$$
\begin{aligned}
\Phi_M(R) &= \frac{\langle M^2 \rangle_R}{3k_B T R^3} \\
&= \frac{\varepsilon - 1}{9\varepsilon(\varepsilon + 2)}\left[(\varepsilon + 2)(2\varepsilon + 1) - 2(\varepsilon - 1)^2\left(\frac{R}{R_c}\right)^3\right]
\end{aligned}
\tag{23}
$$

where k_B is the Boltzmann constant and T the absolute temperature.[46] It contains ε as a parameter. $\Phi_M(R)$, a dimensionless mean square moment fluctuation in units of $3k_B T R^3$, is the result of the simulation, and ε is then obtained by comparison with expression (Equation (23)).

Figure 28 shows the result for the water model known as *soft sticky dipole* (SSD) water which gives good equilibrium and dynamic properties for bulk water.[45] In particular, it results in $\varepsilon = 81$ for bulk water. The plots show good linearity and thus conform well to Equation (23). Slight deviations can be attributed to the small number of molecules and to the inhomogeneity of the medium in smaller cavities. ε can also be read from the intercept of the plots, *i.e.* for $R_c \rightarrow \infty$ for which

$$
\begin{aligned}
\Phi_M^{\infty}(R) &= \frac{(\varepsilon - 1)(2\varepsilon + 1)}{9\varepsilon} \\
&\approx \frac{2\varepsilon}{9} \quad \text{(for } \varepsilon \gg 1\text{)}
\end{aligned}
\tag{24}
$$

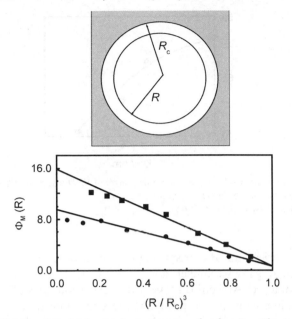

Figure 28 *Definition of model for a spherical cavity of radius R_c within a concentric shell of radius R (upper panel) and simulated values of $\phi_M(R)$ (lower panel), relating to the square dipole moment fluctuation and to the dielectric constant. The calculations are for SSD water in a spherical cavity of 2.44 nm (top curve, giving $\varepsilon = 72$) and 1.22 nm (bottom curve, giving $\varepsilon = 42$).*
(Reprinted with permission from Ref. 45. Copyright (2001) American Chemical Society)

The results of these extrapolations are displayed in Figure 29. It illustrates that the dielectric constant of water decreases with decreasing cavity diameter and drops by a factor of 2 for water in cavities with a diameter of 1.22 nm. This shows that the ability of a medium to respond to an external electrical field is reduced and that the range of Coulomb interactions is increased in pores. Since the cavity walls are non-reactive, the effect originates purely from confinement. It indicates an increasing orientational order of the molecules near the surface of a droplet or the interface towards a non-reactive wall which reduces the ability of these molecules to reorient under the influence of an electric field.

The dielectric constant as discussed above for pore-confined water represents a static limit of a frequency-dependent property. It is also interesting to analyse the time scale of the response of the molecular orientation to fluctuating fields. Such investigations were conducted by Kremer *et al.*[47] using broad band dielectric spectroscopy (10^{-2}–10^{9} Hz) on ethylene glycol and propylene glycol in zeolite nanopores. These hydrogen bonding molecules form glasses which display cooperative dynamics with a temperature-dependent relaxation rate that is normally described in the bulk phase by the *Vogel–Fulcher–Tammann relationship*

$$\frac{1}{\tau} = A \exp\left(\frac{DT_0}{T - T_0}\right) \qquad (25)$$

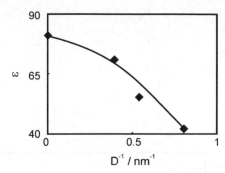

Figure 29   *Values of the dielectric constant of confined SSD water in spherical cavities as a function of the inverse diameter.*
(Data drawn from Ref. 45)

where τ is the reorientational correlation time (the inverse relaxation rate), A is a prefactor, D the fragility parameter, and T_0 the Vogel temperature.[48–50]

Figure 30 compares Arrhenius plots of the relaxation rates of ethylene glycol confined in zeolites with various pore diameters and pore architecture with that of the bulk glass-forming liquid. Clearly, the dynamics in zeolite beta and in the aluminophosphate $AlPO_4$-5 is quite indistinguishable from that in the bulk, and is fitted well with Equation (25), resulting in $A = 6.8 \times 10^{13}$ Hz, $D = -14.8$ and $T_0 = 107.8$ K. In contrast, the same molecule in sodalite shows clean Arrhenius type relaxation with an activation energy of 26 kJ mol^{-1} which corresponds to the limiting value for bulk ethylene glycol at high frequencies and high temperature. The behaviour in silicalite is intermediate and represented by an activation energy of 35 kJ mol^{-1} which is in between the value in silicalite and the apparent value (slope of tangent) for the bulk.

Inspection of the pore structures helps to understand the different behaviour. The silica version of sodalite (pure SiO_2) consists of identical cages with a free inner diameter of 0.6 nm. Ethylene glycol becomes occluded as a structure-directing agent during synthesis with a single molecule per cage and cannot escape unless it is thermally decomposed. Silicalite I is a different modification of SiO_2 with a two-dimensional pore structure of two slightly different elliptical channels with cross sections of 0.56 nm × 0.53 nm and 0.55 nm × 0.51 nm. Zeolite beta is an aluminosilicate with a Si/Al atomic ratio of 40 and a three-dimensional pore system of 0.76 nm × 0.64 nm and 0.55 nm. $AlPO_4$-5 consists of arrays of cylindrical one-dimensional channels with 0.73 nm diameter. The surprising conclusion is that at low temperature the dynamics is fastest in the smallest cage. The single-molecule relaxation of ethylene glycol in sodalite at $T \approx 155$ K is faster by about six orders of magnitude compared with the bulk liquid. It demonstrates that the effect that controls dynamics originates in cooperative interaction of the molecules rather than in spatial constraint by the lattice. This explanation is further corroborated by an analysis of the coordination number based on computer simulations of completely loaded host-guest systems using three different force fields.[47] The result which is

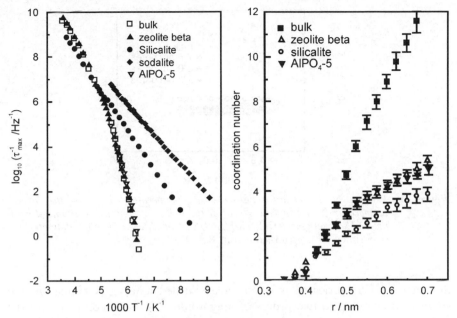

Figure 30 *Left: Relaxation rate versus inverse temperature for ethylene glycol confined in different zeolite host systems. Right: Average coordination number within a cut-off radius r obtained from simulations of ethylene glycol in the bulk liquid and confined in zeolite pores.*
(Reprinted with permission from Ref. 47)

displayed in Figure 30 as a function of cut-off radius r shows clear differences. For $r = 0.66$ nm a coordination number of 11 ± 1 is found for the bulk liquid, which corresponds within error to the value for close packing spheres. For the same cut-off radius in zeolite beta and in $AlPO_4$-5 ethylene glycol has only five neighbouring molecules. One thus has to conclude that an ensemble of only six molecules is sufficient to perform liquid-like dynamics. A further reduction in channel size as in the case of silicalite decreases the average number of neighbouring molecules by about one. This results in a sharp transition from the liquid-like Vogel–Fulcher–Tammann dynamics to single molecule behaviour.

For glass-forming liquids, the product of temperature T and dielectric strength $\Delta\varepsilon$ increases with decreasing temperature, while it should be constant for the case of simple *Debye relaxation*. Propylene glycol displays the expected glassy behaviour in the bulk, but in nanoporous sol–gel glasses with a narrow pore size distribution the character changes to Debye-like character (Figure 31). The fact that the dielectric strength becomes smaller with decreasing pore diameter was attributed to the change in the surface-to-volume ratio of the nanoporous system.[47] This experimental observation parallels the behaviour predicted in Figure 29 for water, but it occurs at significantly larger pore size.

It is also interesting to note that MD simulations of the *quasi*-van der Waals liquid salol (phenyl salicylate) confined in nanopores of 2.5, 5.0 and 7.5 nm

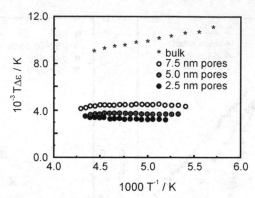

Figure 31 *Product of dielectric strength Δε and temperature T versus the inverse temperature for propylene glycol in the bulk and confined in silanised commercial sol–gel glass (Geltech Inc., USA) of different nominal diameter.* (Reprinted with permission from Ref. 47)

diameter with a lubricated inner surface revealed a faster dynamics in the smaller pores by up to two orders of magnitude.[47] Quite in contrast with the H-bond-forming glycols, one observes over a wide temperature range a dynamics that is fully compatible with that of a bulk liquid. However, with decreasing temperature the Vogel–Fulcher–Tammann dynamics turns suddenly into Arrhenius-type dependence. The transition temperature, as also supported in calorimetric measurements, depends strongly on the pore diameter and allows an estimate of the length scale over which cooperative rearrangements in salol take place.

7.2 Diffusion and Viscosity under Confinement

Diffusion under confinement shows a number of important specialities and surprises in comparison to bulk diffusion. Adsorption, pore architecture, heat transfer and barriers at grain boundaries play an eminent role and have led to considerable confusion because of the discrepancies between the diffusion coefficients, which were determined by different methods and under different experimental conditions. Systematic studies over the last two decades have resolved many of the problems. A detailed and competent overview has been given by Kärger and Ruthven.[51]

The first point that can puzzle initially is the loading dependence of diffusivity in porous materials. The diffusion coefficient can decrease as well as increase or stay relatively independent of loading, dependent on the nature of the porous solid and of the diffusing molecule (Figure 32). A decrease of the mobility with increasing concentration as observed for *n*-hexane in NaX may intuitively be understood by mutual hindrance of the molecules. An increase with concentration can be related to strong adsorption in the pores at sites, which become saturated at higher concentrations, so that the fraction of actually mobile molecules increases with concentration.[52] The two dependencies can balance each other and lead to an apparent concentration-independent diffusion under certain conditions.

The transport diffusivity is commonly written as

$$D_T = \frac{\partial \ln p}{\partial \ln c} D_0 \tag{26}$$

D_0 is a self-diffusion coefficient that describes the molecular mobility in the absence of interaction with the host. The first term, the so-called thermodynamic factor, describes the necessary correction which deviates from unity as soon as the adsorption isotherm deviates from proportionality between sorbate pressure and overall sorbate concentration. Figure 32 demonstrates that this factor can cover as much as two orders of magnitude.

In a homogeneous medium the mean square displacement which describes transport of a diffusing particle in a specified direction with time obeys the Einstein relation

$$\langle x^2(t) \rangle = 2Dt \tag{27}$$

In order for this to be true the probability of a step in positive and negative x-direction must be the same. However, in a one-dimensional pore which is sufficiently narrow so that molecules cannot bypass each other, this probability will depend on the available space. In a filled pore there is no net transport unless all molecules in the pore move at the same time and in the same direction. The situation is analogous to that in a filled theatre where a number of people have to leave the row before a late-comer can get to his seat in the

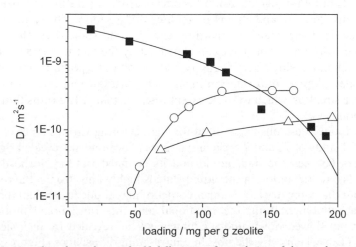

Figure 32 *Loading dependence of self-diffusivities for a choice of three substrates in NaY zeolite observed by pulsed field gradient NMR. Squares: n-hexane at 358 K, circles: ammonia at 293 K, triangles: water at 393 K.*
(Drawn using data from Refs. 52 and 53)

row. Incompletely filled pores leave more freedom, but the time-dependence for the mean square displacement is different from that of normal diffusion (Equation (27)) and assumes

$$\langle x^2(t) \rangle = 2Ft^{1/2} \tag{28}$$

Here, F is the mobility factor which is proportional to a Langmuir-type occupation probability of a pore site, $\theta/(1-\theta)$.[52] This type of restricted diffusion in a one-dimensional pore is called *single file diffusion*.

The adsorption equilibrium is normally dominated by a negative heat of adsorption so that in the presence of a porous solid a significant fraction of gaseous molecules will enter the pores to reach the favoured adsorbed state. This also means that the molecules experience an energy barrier at the surface of a crystallite which prevents them from leaving the pore. The molecule gets reflected at the crystallite surface, and the displacement trajectory explores the shape of a crystallite much more intensively than the intercrystalline space. Diffusion is therefore not homogeneous. Eventually, with increasing temperature, a substantial fraction of molecules is able to surpass the barrier and enter the gas phase in the surrounding intercrystalline space.

When the gas pressure is sufficiently low in the intercrystalline voids, which normally means low temperatures, the molecules collide only with nearby crystallite surfaces and undergo transport by multiple reflections. This situation is known as *Knudsen diffusion*. However, when the gas pressure increases upon increasing temperature the collisions will mostly be with other gas phase molecules. The increase of the thermodynamic factor is thus compensated to a large extent by the decrease of the mean free path due to pressure, so that the coefficients of long-range diffusion become nearly temperature independent.[54]

Normally, long alkane chains in zeolites diffuse more slowly than shorter ones, due to their higher activation energy, but sometimes the opposite behaviour is found.[55] This has been explained by the fact that short molecules may fit well into a single zeolite cavity and adopt a low potential energy, while longer ones adopt a conformation in which at least one end extends through a window, a situation which to some extent resembles already a transition state of a diffusion step.

NMR is a well-established method for characterising various types of MD. A ^{1}H and ^{2}H study of T_1 and T_2 relaxation times of acetonitrile revealed significant differences between the bulk liquid and its confined state in Sorbosil porous silica.[56] The reorientation in the liquid state is highly anisotropic, the reorientation about the symmetry axis being an order of magnitude faster than the overall molecular tumbling. Both, the rotational and the translational motions are considerably slowed down under confinement, as revealed by their activation energies (Table 1) and by the absolute intra-grain diffusivities (Figure 33). It is believed that the interactions between the surface hydroxyl groups and the CN nitrogen of the probe molecule hinder the molecular tumbling motion considerably. The lower self-diffusion coefficients under confinement may be ascribed to

Table 1   *Activation energies derived from NMR measurements for rotational and translational motion of acetonitrile in the bulk and confined in porous silica (Data from Ref. 56. Reproduced with permission from the PCCP Owner Societies)*

| | Activation energy/kJ mol^{-1} | |
Sample	Anisotropic rotation	Self-diffusion
Bulk	7.5	12
20 nm confinement	9.4	20
6 nm confinement	9.7	27

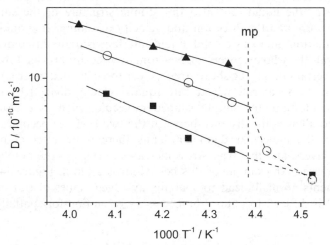

Figure 33   *Intra-grain diffusivities of acetonitrile confined in 6 nm silica (squares), 20 nm silica (circles) and bulk acetonitrile (triangles). mp indicates the melting point of bulk acetonitrile.*
(Data selected from Ref. 56. Reproduced with permission from the PCCP Owner Societies)

an increased effective viscosity. There is a discontinuity near the bulk melting point in the 20 nm but not in the 6 nm pore diameter silica, giving evidence that the melting point decreases in narrow pore material.

Equilibrium and non-equilibrium MD simulations of a KCl electrolyte in cylindrical nanopores of 0.475–1.58 nm radius revealed a decrease of hydrogen bonding, diffusion coefficients and ion conductivity with decreasing pore radius.[57]

Transport properties of nanosystems with dimensions below about 10 molecular diameters differ dramatically from those specific to bulk systems. It is found that molecules become progressively more ordered, and that their mobility sharply decreases. This ordering is further modified by shear. The liquid density across molecularly thin films is not uniform but has an oscillatory profile with a periodicity close to the molecular diameter. The thinner the layer the sharper these density oscillations become. The order near a surface is analogous to the well-known radial distribution function that describes the average order around an atom or molecule in an isotropic liquid. When the film thickness is only a few molecular diameters, the order in molecular layers which are oriented parallel to the surface is so strict that it becomes difficult to distinguish a liquid from a solid.[58,59] In a direction parallel to the surface the order is less well defined.

Qualitatively, the *effective viscosity* in molecularly thin films between two molecularly smooth surfaces can be 10^5–10^7 times the bulk value, and molecular relaxation can be 10^{10} times slower.[58] These properties depend not only on the nature of the liquid, but also the atomic structure of the surfaces, the normal pressure and the direction and velocity of sliding. It should be noted that the situation on a single solid–liquid interface is quite different.

In general, the whole concept of Newtonian viscosity breaks down for thin films. In a certain range, the shear stress (shear force divided by area) no longer depends on the shear rate (shear rate equals velocity divided by film thickness).[58] In addition, for simply shaped molecules such as cyclohexane or unbranched alkanes the distance between the two surfaces becomes *quantised* in units of the molecular diameter. Thus there is an integral number of molecular layers, and the shear stress increases in steps as the number of such layers decreases. An example of this behaviour is given in Figure 34.

Experiments normally lead to spatially averaged values of transport properties, as it is difficult to obtain the necessary resolution for spatially resolved

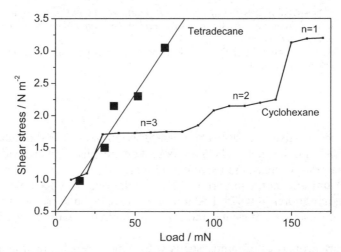

Figure 34 *Variation of shear stress with load in the steady state for films in the solid-like state.*
(Selected data reprinted with permission from Ref. 58. Copyright (1990) American Institute of Physics)

images of these phenomena on a nanometer scale. Theoretical methods are superior in this respect. In particular, various types of MD simulations have led to amazing insight in nanoscale properties.

One of the transport properties of confined nanofluids which have been of high interest is viscosity. It has been addressed for nanofluids confined in slit pores using non-equilibrium statistical mechanics approach by Pozhar *et al.*[60] and compared with non-equilibrium MD simulations. The bulk viscosity of a Newtonian fluid in a macroscopic slit pore is constant across the pore, as is the density, and the velocity profile across the pore is parabolic with a value equal to zero at the pore walls. In contrast, in a nanoscopic slit pore the interaction of the fluid molecules with the walls leads to a modulation of the density, and in sufficiently narrow pores it results in layers which may be regarded as separate phases (see Chapter 7 Section 4 and Figure 24). The velocity profile across the pore is predicted by theory to have a remarkably similar appearance as the density profile, showing maxima where the density has a maximum, and for narrow pores it is far from parabolic. Also the viscosity profile has a similar appearance, in particular near the walls, whereas it is somewhat washed out near the pore centre.[60]

Of interest for engineering purposes is the averaged value of the viscosity across the pore. It is plotted in Figure 35 for slit pores of different dimensions for a moderately dense nanofluid. We see that it moves increasingly to higher values as the pore gets narrower, and it reaches values several times larger than the bulk value under conditions which are otherwise the same.

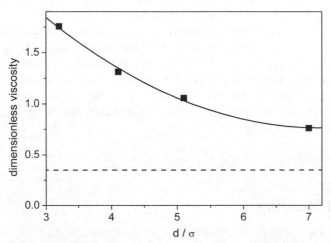

Figure 35 *Pore average dimensionless viscosity from a non-equilibrium statistical mechanics approach for a nanofluid confined in slit nanopores of various width d in units of the molecular diameter parameter σ. The broken line gives the viscosity for the bulk fluid at the same density $<n\ \sigma^3> = 0.442$ and temperature $k_BT/\varepsilon = 0.729$.*
(Data drawn from Ref. 60)

Key Points

- Narrow pores reduce the maximum coordination number that an adsorbed system can adopt. This usually results in marked depressions of melting points, triple points and critical points.
- Fluids adsorbed on surfaces and in narrow pores exhibit discrete layer phases with new critical points and new triple points.
- Hysteresis loops in adsorption isotherms of gases in mesopores reflect non-equilibrium situations due to the presence of side minima of the free enthalpy as a function of adsorbate pressure.
- The narrower a pore the lower the melting point of a wetting adsorbate and the broader the calorimetric melting peak. The boiling point of the same system increases above its bulk value.
- The freezing point of a wetting liquid layer inside a pore increases at the inner surface and decreases at the outer surface because of the opposite sign of curvature.
- For monodisperse pores the width of a melting peak reflects the cooperative nature of a phase transition. In the thermodynamic limit (*i.e.* for an infinite number of particles) the peak approaches a delta function.
- The surface tension of a cluster exerts a pressure which may force the cluster to adopt a high pressure phase at small diameter.
- The dielectric constant of water decreases with decreasing pore size.
- The viscosity of a nanofluid increases with decreasing pore size.
- The dynamics of a hydrogen bonding liquid can be much faster in a narrow pore where the molecule is isolated and can move freely without having to break hydrogen bonds to its neighbours.
- Diffusion coefficients of molecules in pores decrease with increasing loading due to mutual hindrance of diffusive motion. However, they may also increase with loading if the first molecules saturate strong adsorption sites so that further molecules see a 'lubricated' surface.
- For confinement between flat surfaces at small distances, the shear stress is quantised and larger for unimolecular than for bimolecular or trimolecular layers.

General Reading

- L.D. Gelb, K.E. Gubbins, R. Radhakrishnan and M. Sliwinska-Bartkowiak, Phase separation in confined systems, *Rep. Prog. Phys.*, 1999, **62**, 1573.
- M. Thommes, R. Köhn and M. Fröba, *J. Phys. Chem. B*, 2000, **104**, 7932.
- F. Rouquerol, J. Rouquerol and K. Sing, *Adsorption by Powders and Porous Solids*, Academic Press, San Diego, 1999.

- K.-J. Hanszen, Theoretische Untersuchungen über den Schmelzpunkt kleiner Kügelchen, Z. *Physik*, 1960, **157**, 523.
- J. Kärger and D.M. Ruthven, *Diffusion in Zeolites and Other Microporous Solids*, Wiley, New York, 1992.
- J. Kärger and H. Pfeifer, NMR self-diffusion studies in zeolite science and technology; *Zeolites*, 1987, **7**, 90.

References

1. S. Brunauer, P.H. Emmett and E. Teller, *J. Am. Chem. Soc.*, 1938, **60**, 1723.
2. A.W. Adamson, *Physical Chemistry of Surfaces*, Wiley, Toronto, 1982.
3. S.J. Gregg and K.S.W. Sing, *Adsorption, Surface Area and Porosity*, Academic Press, London, 1982.
4. G. Horvath and K. Kawazoe, *J. Chem. Eng. Japan*, 1983, **16**, 470.
5. E.P. Barrett, L.G. Joyner and P.P. Halenda, *J. Am. Chem. Soc.*, 1951 **73**, 373.
6. L.D. Gelb and K.E. Gubbins, *Langmuir*, 1999, **15**, 305.
7. M.P. Rossi, H. Ye, Y. Gogotsi, S. Babu, P. Ndungu and J.-C. Bradley, *Nano Lett.*, 2004, **4**, 989.
8. F. Rouquerol, J. Rouquerol and K. Sing, *Adsorption by Powders and Porous Solids*, Academic Press, New York, 1998.
9. M. Thommes, R. Köhn and M. Fröba, *J. Phys. Chem. B*, 2000, **104**, 7932.
10. R.J.-M. Pellenq, B. Rousseau and P.E. Levitz, *Phys. Chem. Chem. Phys.*, 2001, **3**, 1207.
11. L.D. Gelb, K.E. Gubbins, R. Radhakrishnan and M. Sliwinska-Bartkowiak, *Rep. Prog Phys.*, 1999, **62**, 1573.
12. M. Fisher and H. Nakanishi, *J. Chem. Phys.*, 1981, **75**, 5857.
13. K. Morishige and M. Shikimi, *J. Chem. Phys.*, 1998, **108**, 7821.
14. K. Morishige, H. Uematsu and N. Tateishi, *J. Phys. Chem. B*, 2004, **108**, 7241.
15. L.H. Cohen, *J. Am. Chem. Soc.*, 1938, **60**, 433.
16. B. Coasne, A. Grosman, C. Ortega and M. Simon, *Phys. Rev. Lett.*, 2002, **88**, 256102.
17. J.A. Duffy, N.J. Wilkinson, H.M. Fretwell, M.A. Alam and R. Evans, *J. Phys.: Condens. Matter*, 1995, **7**, L713.
18. D.W. Brown, P.E. Sokol, A.P. Clark, M.A. Alam and W.J. Nuttall, *J. Phys.: Condens. Matter*, 1997, **9**, 7317.
19. G. Zhao, B. Gross, H. Dilger and E. Roduner, *Phys. Chem. Chem. Phys.*, 2002, **4**, 974.
20. A. Schreiber, I. Ketelsen and G.H. Findenegg, *Phys. Chem. Chem. Phys.*, 2001, **3**, 1185.
21. P.Z. Pawlow, *Z. Phys. Chem.*, 1909, 65, 1; and *ibid*. 545.
22. M. Miyahara and K.E. Gubbins, *J. Chem. Phys.*, 1997, **106**, 2865.
23. H. Reiss and I.B. Wilson, *J. Colloid Sci.*, 1948, **3**, 551.

24. K.-J. Hanszen, *Z. Phys.*, 1960, **157**, 36.
25. J.R. Sambles, *Proc. R.. Soc. London, A*, 1971, **13**, 324; and *ibid.* 339.
26. Y.G. Chushak and L.S. Bartell, *J. Phys. Chem. B*, 2001, **105**, 11605.
27. G.N. Antonov, *J. Chim. Phys.*, 1905, **5**, 372.
28. W.R. Tyson and W.A. Miller, *Surf. Sci.*, 1977, **62**, 267.
29. K.M. Unruh, T.E. Huber and C.A. Huber, *Phys. Rev. B*, 1993, **48**, 9021.
30. A.N. Goldstein, C.M. Echer and A.P. Alvisiatos, *Science*, 1992, **256**, 1425.
31. J. Woltersdorf, A.S. Nepijko and E. Pippel, *Surf. Sci.*, 1981, **106**, 64.
32. C. Solliard and M. Flueli, *Surf. Sci.*, 1985, **156**, 487.
33. J.H. Strange, M. Rahman and E.G. Smith, *Phys. Rev. Lett.*, 1993, **71**, 3589.
34. J.Z. Larese, *Acc. Chem. Res.*, 1993, **26**, 353.
35. M. Sliwinska-Bartkowiak, J. Gras, R. Sikorski, R. Radhakrishnan, L. Gelb and K.E. Gubbins, *Langmuir*, 1999, **15**, 6060.
36. R. Radhakrishnan and K.E. Gubbins, *Mol. Phys.*, 1999, **96**, 1249.
37. A. Huerta, O. Pizio and S. Sokolowski, *J. Chem. Phys.*, 2000, **112**, 4286.
38. I. Brovchenko, A. Geiger and A. Oleinikova, *Phys. Chem. Chem. Phys.*, 2001, **3**, 1567.
39. L.V. Entov, V.A. Levchenko and V.P. Voronov, *Int. J. Thermophys.*, 1993, **14**, 221.
40. V.P. Voronov and V.M. Buleiko, *J. Exp. Theor. Phys.*, 1998, **86**, 586.
41. T. Okubo, H. Kanoh and K. Kaneko, *Aust. J. Chem.*, 2003, **56**, 1013.
42. A.P. Alivisatos, *J. Phys. Chem.*, 1996, **100**, 13226.
43. C.Y. Zhang, C.X. Wang, Y.H. Yang and G.W. Yang, *J. Phys. Chem. B*, 2004, **108**, 2589.
44. C.X. Wang, Y.H. Yang, Q.X. Liu and G.W. Wang, *J. Phys. Chem. B*, 2004, **108**, 728.
45. S. Senapathi and A. Chandra, *J. Phys. Chem. B*, 2001, **105**, 5106.
46. D.J. Berendsen, Molecular dynamics and Monte Carlo calculations on water, *CECAM Report*, 1972.
47. F. Kremer, A. Huwe, M. Arndt, P. Behrens and W. Schwieger, *J. Phys.: Condens. Matter*, 1999, **11**, A175.
48. H. Vogel, *Phys. Z.*, 1921, **22**, 645.
49. G.S. Fulcher, *J. Am. Chem. Soc.*, 1925, **8**, 339.
50. G. Tammann and G. Hesse, *Anorg. Allgem. Chem.*, 1926, **156**, 245.
51. J. Kärger and D.M. Ruthven, *Diffusion in Zeolites and Other Microporous Solids*, Wiley, New York, 1992.
52. J. Kärger, Diffusion under confinement, in *Sitzungsberichte der Sächsischen Akademie der Wissenschaften zu Leipzig, Mathematisch-Naturwissenschaftliche Klasse*, Band 128, Heft 6, Verlag Sächsische Akademie der Wissenschaften, Leipzig, 2003.
53. J. Kärger and H. Pfeifer, *Zeolites*, 1987, **7**, 90.
54. J. Kärger, F. Stallmach and S. Vasenko, *Magn. Reson. Imaging*, 2003 **21**, 185.
55. H. Jobic, A. Méthivier, G. Ehlers, B. Farago and W. Haeussler, *Angew. Chem.*, 2004, **116**, 116.

56. D.W. Aksnes, L. Gjerdäker, L. Kimtys and K. Førland, *Phys. Chem. Chem. Phys.*, 2003, **5**, 2680.
57. Y.W. Tang, K.-Yu. Chan and I. Szalai, *J. Phys. Chem. B*, 2004, **108**, 18204.
58. M.L. Gee, P.M. McGuggian, J.N. Israelachvili and A.M. Homola, *J. Chem. Phys.*, 1990, **93**, 1895.
59. S. Granick, *Science*, 1991, **253**, 1374.
60. L.A. Pozhar, E.P. Kontar and M.Z.-C. Hu, *J. Nanosci. Nanotech.*, 2002 **2**, 209.

CHAPTER 8
Nucleation, Phase Transitions and Dynamics of Clusters

1 Melting Point and Melting Enthalpy

1.1 Introduction

Concepts such as temperature, phase and melting were originally defined for macroscopic systems. Meanwhile, one has learned to generalise these concepts to make them applicable to small systems. As regards melting, there is the question about a reliable experimental signature. The traditional criterion is a calorimetric observation of a latent heat of melting. Alternatively, one can look for the disappearance of a Debye–Scherrer-like pattern in transmission electron microscopy (TEM) diffraction experiments, but one has to be cautious about local heating by the electron beam. In simulations, one often uses the empirical Lindemann criterion which states that the thermal fluctuations of the internuclear distances become larger than 10–15% when bulk matter melts.

For solid-to-liquid transitions of finite systems, one finds four main differences with respect to bulk behaviour:[1]

- There is a general size-dependent depression of the melting point.
- The latent heat per atom or per molecule is smaller.
- The transition is not sharp but spread over a finite temperature range.
- The heat capacity of an isolated system can become negative, *i.e.* the temperature of a system can decrease upon addition of energy.

In the second half of the past century size-dependent melting of particles created great interest, and several phenomenological models addressing the melting point depression of small metal particles were developed (see Chapter 7, Section 3). The latent heat of melting, ΔH_m, was normally assumed to adopt its bulk value, since it was essentially impossible to determine it independently of other parameters such as cluster radius or surface tensions. With the availability of improved quantum chemical simulation programmes the question was addressed and the result suggested a size dependence of ΔH_m. On the experimental side, TEM proved suitable for studying size-dependent melting, but the heat associated with this process was still not accessible. A detailed early

209

investigation on size-dependent melting was conducted by Buffat and Borel[2] using electron diffraction of gold particles formed by condensation of gold vapour on a thin amorphous carbon film of low affinity for gold. It was found that for particles of 6 nm diameter, the melting point exhibits a dramatic drop to *ca.* 310 K, which is less than 25% of its bulk value of 1336 K. Depending on the model on which the analysis was based, the authors were compelled to assume a solid-core–liquid-shell model with a shell thickness of about 0.6 nm. Later work revised these extremely low melting points and reported size- and temperature-dependent structural transitions just below the melting point of gold nanoparticles.[3] The reason for the significant discrepancy between the two sets of data is not quite clear.

1.2   Supported Tin Cluster

The obvious technique for determination of melting points and of latent heats of melting is calorimetry, but instruments that could deal with particles of nanometre dimensions and with heats corresponding to a few tens of meV (the bulk melting enthalpy of tin amounts to 73 meV per atom) had to be developed first. The main requirement for sufficient (nano-Joule) sensitivity consists in a comparably low heat capacity of the essential parts of the calorimeter. Advanced thin film and membrane fabrication technology afforded a dramatic reduction of the calorimeter mass, and in the mid-1990s Lai *et al.*[4] reported the first direct determination of latent heats of fusion and melting points of vapour-deposited tin nanoparticles with radii ranging from 5–50 nm, along with scanning electron microscopic measurements of the cluster size. The key results of this report are displayed in Figures 1 and 2.

Figure 1 shows that the melting point of tin particles with a radius of 5 nm is below that of bulk Sn (232 °C) by about 70 °C. At the same time, ΔH_m drops in a similar fashion from its bulk value of 58.9 J g^{-1} to as little as one–third of it.

The results were analysed in terms of the solid-core–liquid-shell model by Hanszen (see Chapter 7, Section 3, Equation 19), using the interfacial surface tension γ_{sl} and the thickness t_0 of the liquid surface layer as adjustable parameters. This resulted in $\gamma_{sl}=48$ mN m^{-1}, in agreement with macroscopic determinations for a planar surface, and in $t_0=1.8$ nm. Both were assumed to be size independent. The depressed melting point was ascribed to the solid core. For the analysis of the latent heats, it was assumed that ΔH_m of the solid core corresponds to its bulk value. Since the surface layer is already liquid, the measured *average* latent heat per atom decreases according to the volume fraction of the solid core:

$$\Delta H_m = \Delta H_0 \left(1 - \frac{t_0}{r}\right)^3 \qquad (1)$$

Fitting the experimental data with t_0 as a single free parameter yields $t_0=1.6$ nm, which agrees well with the corresponding value that was determined independently from the melting point data (Figures 1 and 2).

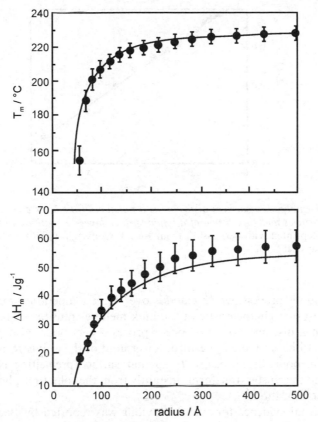

Figure 1 *Size dependence of melting point (top) and latent heat of fusion per atom (bottom) of tin particles.*
(Reprinted with permission from Ref. 4. Copyright (1996) by the American Physical Society)

The analyses by Lai *et al.* lead to discussions about the consequences and limitations of the model in several respects.[5,6] First, according to Young's equation, a wetting liquid layer is only stable when it leads to a reduction in surface free energy, that is when the contact angle θ is <90°, that is when $\gamma_{sv} > \gamma_{sl}$. The thickness t_0 of the liquid layer is determined by balancing two opposing effects: the gain of surface free energy when the liquid film is formed and the melting free energy spent for the generation of the liquid layer. The thickness t_0 turns out to be in the order of several monolayers only. While the surface free energy term is to a good approximation constant, the free energy difference between solid and liquid is primarily of entropic nature and therefore proportional to a temperature offset. For a sufficiently large cluster, the model implies that there is a critical temperature T_c where surface melting is initiated. Increasing the temperature from T_c causes the thickness t_0 of the liquid layer to increase continuously until the melting point of the solid cluster core is reached.

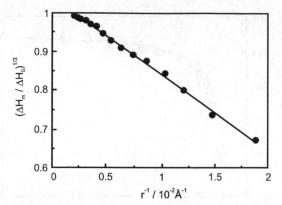

Figure 2 *Normalized latent heat of fusion as a function of the inverse core radius of tin particles under exclusion of a liquid surface layer (Equation (1)).*
(Reprinted with permission from Ref. 4. Copyright (1996) by the American Physical Society)

This continuous process, which extends over some temperature range, smears out the first-order characteristic of the bulk melting transition. Melting of the core is then a discontinuous first-order process. The model also predicts the puzzling feature that there is a critical lower limit of the radius R_c at which the core melting point drops below T_c so that surface premelting is no longer possible, and the cluster transforms instantly from the solid to the liquid state in a first-order transition.[5]

Experimental evidence for such behaviour was reported by Bachels *et al.*[5] (Figures 3 and 4), based mostly on the above results by Lai *et al.*[4] Figure 3 indicates that the melting point of a free cluster with close to 500 tin atoms may deviate from the predicted monotonic behaviour. This was taken as clear evidence for a change in the melting mechanism. Furthermore, the same cluster has a latent heat of melting that is significantly higher than that of much larger supported clusters (Figure 4). These observations are in line with independent work by Shvartsburg and Jarrold[7] who observed no signature of melting for tin cluster ions containing 10–50 atoms, and they concluded that the melting point is at least 50 K above that of bulk tin. The authors considered heavy reconstruction of the surface as a possible origin of the unexpected finding. An analogous finding was also reported for Ge_{39}^+ and Ge_{40}^+ clusters which were found to melt around 550 K, substantially *above* the 303 K melting point of bulk gallium and in conflict with expectations based on scaling laws that these clusters will remain liquid below 150 K, while their heats of fusion per atom are only about 20% of the bulk values.[8] No melting was detected for Ga_{17}^+ up to 720 K. Density functional molecular dynamics calculations verified extremely high melting temperatures as revealed by a maximum in specific heat that occurred near 650 K for Ga_{17}^+ and as high as 1400 K for Ga_{13}^+.[9] This was attributed to the formation of mainly covalent bonding concomitant with ring-like structures, in contrast with the covalent-metallic bonding in the bulk.

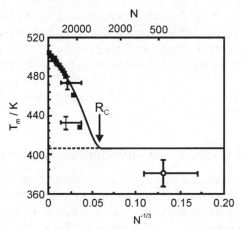

Figure 3 *Melting point of supported (closed symbols[4]) and isolated (open symbol) tin nanoclusters as a function of the inverse cluster radius. The solid line is a prediction of the size-dependent T_m based on the liquid-shell model, and the dashed line corresponds to the critical temperature T_c below which the model predicts no surface premelting. The location of the open symbol is taken as clear evidence of a change in the size dependence.*
(Reprinted with permission from Ref. 5. Copyright (1996) by the American Physical Society)

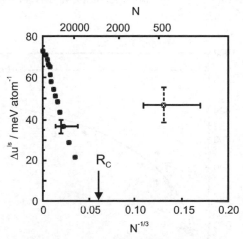

Figure 4 *Latent heat of fusion of supported (closed symbols[4]) and isolated (open symbol) tin nanoclusters as a function of the inverse cluster radius. The liquid-shell model predicts that the latent heat of melting vanishes at a critical radius R_c. The enhanced latent heat of melting for $R < R_c$ (open symbol) was taken as evidence for a change in the melting mechanism.*
(Reprinted with permission from Ref. 5. Copyright (1996) by the American Physical Society)

The above results were disputed by Kofman *et al.*[6] who insisted that the experiments on supported clusters by Lai and on free clusters by Bachels are not strictly comparable. Furthermore, they doubted that a phenomenological model that is based on classical thermodynamics is appropriate to adequately describe the behaviour of such small clusters. Instead, they suggested the existence of a dynamical coexistence of different phases. We will return to such an interpretation further below.

1.3 Melting of Capped Cadmium Sulfide Nanocrystals

A spectacular decrease in the melting point by about 1100 K relative to its bulk value of 1678 K was observed by the disappearance of electron diffraction patterns or a change in the dark field for thiophenol-(or mercaptoacetic acid)-capped CdS nanoparticles of 1.2–7.6 nm radii (Figure 5).[10]

1.4 Free Sodium Clusters

A set of very well-defined experiments on the melting of size-selected *free* sodium clusters was performed by the group of Haberland in Freiburg. An elegant method consisting of photofragmentation of mass-selected clusters with a well-defined temperature was developed. The appearance of certain photo-fragments is a function of the internal energy. In this way, the cluster itself serves as a calorimeter.[12–14] It was the first experimental confirmation for the predicted coexistence of liquid and solid clusters over a significant temperature range. Furthermore, and perhaps unexpectedly, it was shown that the melting

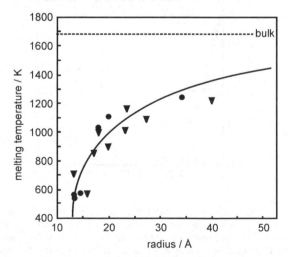

Figure 5 *Melting temperature versus size for CdS nanocrystals.*
(Reprinted with permission from Ref. 11. Copyright (1996) American Chemical Society)

temperature and the latent heat depend sensitively and in a pronounced non-monotonic way on the cluster size.

Figure 6 displays the internal energy of a cationic cluster of 192 sodium atoms, along with its bulk counterpart scaled to 192 atoms and its derivative with respect to temperature – the heat capacity. Three important pieces of information are obtained from this figure: (i) The melting point, read at the maximum of the heat capacity curve, is lowered by *ca.* 130 K relative to its bulk value of 371 K; (ii) the latent heat of melting *q* is only about half of its bulk value; and (iii) melting extends over a range of about 50 K.

The size dependence of the melting point is displayed in Figure 7. The first and expected observation is that the melting temperatures are about 30% below the bulk value of 371 K.[15] More interestingly, the melting point fluctuates considerably with cluster size. It should be noted that high melting points do not coincide with clusters of particular stability. Neither geometry-based magic numbers nor shell-closing effects match the observed maxima in the curve, as indicated by dashed and dotted lines, respectively. However, it was shown more recently that the maxima in the *energies and entropies of melting* which appear to be more fundamental than the melting point correlate with geometric shells

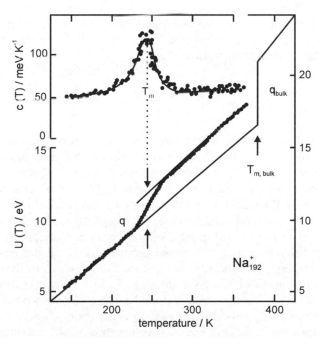

Figure 6 *Heat capacity c(T) and internal energy U(T) of a sodium cluster with 192 atoms and unit positive charge. The solid line gives the scaled caloric curve of bulk sodium, and the dotted line and arrows indicate where the melting point and the latent heat are read off.*
(Reprinted with permission from Ref. 13. http://www.nature.com/)

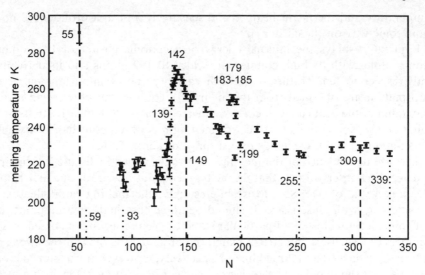

Figure 7 *Melting temperature of sodium clusters as a function of size. The bulk melting temperature is 371 K. Dotted lines give the cluster size at which electronic shells close, dashed lines where geometric shells close.*
(Reproduced with permission from Refs. 1 and 14)

closing.[16] The various parameters are related as follows. For clusters in thermal equilibrium the free energy difference between the solid and the liquid disappears, and the melting temperature obeys

$$T_m = \frac{q}{\Delta s} \tag{2}$$

where q and Δs are the latent heat and the entropy change per atom, respectively, and the two were found to have almost the same size dependence and fluctuation, while T_m changes much more slowly. Since the absolute entropy per atom of the liquid should depend little on size, Δs is expected to be dominated by fluctuations of the entropy of the solid clusters. This was confirmed to be the case, and a simple hard-sphere model shows that the configurational entropy of incomplete surface layers of clusters reproduces the experimental behaviour well.[16] A closed geometric shell cluster has zero configurational entropy, but one additional atom can be put onto the surface (or one hole left in a surface) in a given number of ways, two additional atoms in many more ways, so that the configurational entropy increases on both sides of a closed-shell geometry. Because the size dependences of q and Δs are slightly different in the vicinity of a closed-shell configuration, the maxima in T_m are somewhat shifted with respect to those of q and Δs.

Figure 8 shows the latent heats of melting per atom as a function of cluster size. The main features seem to correlate with those of the melting point curve in Figure 7. This correlation appears physically plausible since the higher the

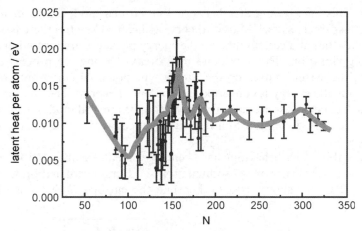

Figure 8 *Latent heat per atom as a function of atoms in sodium clusters. The bulk value amounts to 0.0275 eV. The line is drawn only as a guide to the eye.* (Reproduced with permission from Refs. 1 and 14)

melting point the more the crystal lattice can withstand thermal energy and therefore the higher the latent heat. This correlation also applies to the bulk value of 27.5 meV.

From everyday experience we expect that a system should get warmer when energy is added to it. However, *negative heat capacities* have long been known in astrophysics where energy can be added to a star or star cluster which then cools down.[17] In addition they have been predicted to occur from simulations in the melting of small atomic clusters.[18] This remarkable effect implies that energy is not an extensive quantity, *i.e.* the total energy of a system is not equal to the sum of energies of arbitrary isolated subsystems. This happens when there is interaction between the different subsystems. Macroscopic systems avoid negative heat capacities by phase separation into liquid and solid fractions. Upon melting, the added energy is converted entirely into potential energy, so that the temperature remains constant. A small system may avoid a partially molten state by also converting some of its kinetic energy into potential energy. Therefore, the cluster can become colder, while its total energy increases.

A chemist may understand a negative heat capacity on the basis of the following example: Imagine an isolated S_8 ring-shaped molecule under collision-free conditions. All the energy that exceeds its zero-point energy and thus counts as 'heat' is stored in excitations of internal modes such as stretching and bending vibrations, ring puckering and pseudo-rotation. Initially, the sum of these excitation energies may not be sufficient to break one of the sulfur–sulfur bonds. The molecule can then be heated up by absorbing photons from a radiation field in the infrared or by undergoing inelastic scattering with more events on the Stokes side than on the anti-Stokes side of the exciting wavelength. At some point, the sum of these excitation energies is sufficient to break a bond. This energy migrates between the various modes, and in terms of Rice–Ramsperger–Kassel–Marcus theory of chemical kinetics there is a certain probability that at a certain time all the energy packages are located in a single stretching vibration. Thus the bond may break. During the process of bond breaking the system moves up the Morse-type potential curve until dissociation, *i.e.* kinetic energy is converted into potential energy. The ring-opened S_8 diradical has much less excitation energy left in its internal modes, which means it has cooled off, giving rise to a negative heat capacity in the microcanonical ensemble.

Remarkably, such behaviour has been experimentally observed for the first time for a metal cluster of 147 sodium atoms.[19] For an explanation we refer to Figure 9. Part (a) shows as a broken line the entropy $S(E)$ for a *canonical*

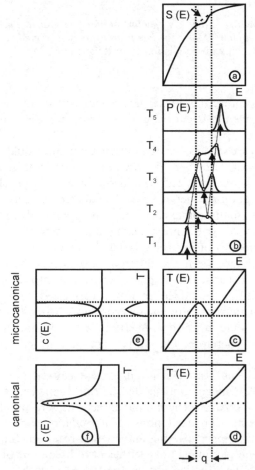

Figure 9   *Entropy $S(E)$ (a), energy distribution $P(E)$ (b), caloric curve $T(E)$ (c and d) for a system with negative heat capacity in the microcanonical (e) but not the canonical (f) ensemble.*
(Reproduced with permission from Ref. 1)

ensemble (a system which has a fixed temperature and a thermal distribution of energies, *e.g.* of a cluster in contact with a heat bath), and superimposed the entropy for the *microcanonical ensemble* (which has a fixed energy, such as an isolated particle in a vacuum). For an infinitely large system, the two ensembles give the same result, but for finite systems there can be differences, in particular near phase transitions, as indicated in the figure by the arrow. In this region *q*, the curvature of $S(E)$, is inverted (colloquially called backbending). Part (b) of the figure gives the canonical energy distribution $P_T(E)$. At the lowest and the highest temperatures, at sufficient distance to the melting range, the distribution is monomodal. However, where the mean energy, denoted by arrows, falls in the inverted region of $S(E)$ the distribution is bimodal. The system never populates the energy levels in the centre of the back-bending region as much as those above and below. In the same range, the temperature of the microcanonical ensemble drops when the energy increases (part (c)), while the canonical ensemble shows a somewhat washed-out normal first-order behaviour of a macroscopic system where the temperature remains constant over the melting range (part (d)). Finally, the derivative of the caloric curve $E(T)$ with respect to temperature is known as the heat capacity $c(T)$. There is nothing unusual in the canonical representation (part (f)) where *c* is always positive, but for the microcanonical system the heat capacity diverges and jumps to negative values over the region *q*.

Several interesting questions are left open at this point. One may ask[1] what other cluster sizes will show this remarkable effect, and how does it depend on the cluster charge (which is singly positive in the present case)? Are there similar effects in other types of phase transitions, for example in magnetic ones? Finally: Are there any applications which take advantage of this behaviour?

1.5 Isolated Silver Clusters

Molecular dynamics simulations revealed three distinctive melting mechanisms in nanoparticles consisting of $N=13$–3871 silver atoms (Figure 10).[20] The melting of large particles in the range $Ag_{258-3871}$ (region I) is explained well by the surface premelting model where the surface layers start to melt at a critical temperature T_c which is below the melting point T_m of the core. The thickness of the liquid layer depends on the local curvature. It increases continuously with temperature until a uniform curvature of the core is attained at T_m. In this region the simulated T_m is described well by the model of Sambles (Equation 19 of Chapter 7), which is based on the experimental bulk melting point T_0 of 1234 K and the bulk melting enthalpy ΔH_m of 106 J g^{-1} (11.4 kJ mol^{-1}), $\gamma_{sl}=184$ mN m^{-1}, $\gamma_{lv}=910$ mN m^{-1}, and a best fit value of 1.01 nm for t_0. The only adjustable parameter was t_0. For particles with $116<N<258$ atoms corresponding to a cluster radius of $r \approx t_0 \approx 1.05$ nm the melting point appears to remain constant at *ca.* 640 K (region II). For smaller clusters T_m shows again a monotonic decrease (region III), but analysis of the root-mean-square fluctuations of the atoms in the different shells reveals that the entire cluster transforms without any evidence of surface premelting. Instead, over a finite temperature range there is a

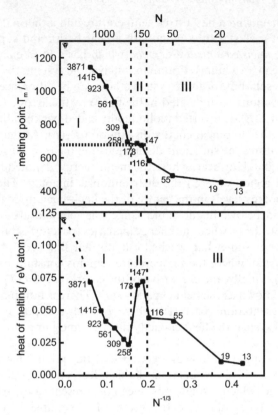

Figure 10   *Melting point and overall latent heat of fusion of isolated silver particles as functions of the inverse particle radius from molecular dynamics simulations (filled squares, with number N of atoms indicated) and bulk experimental values (inverted triangles).*
(Reprinted with permission from Ref. 20. Copyright (2001) American Chemical Society)

dynamic coexistence of a solid-like and a liquid-like form. Each of the atoms swaps back and forth between the two types of behaviour, resembling the inter-conversion of isomers. This *dynamic coexistence melting* is different from mac-roscopic behaviour where two distinct phases are identified and simultaneously present in equilibrium in the same system.

The transition region (II) is also evident as a distinct maximum in the latent heat of melting, and interestingly this occurs at a cluster size of around 147 atoms as observed already in experimental latent heats of sodium clusters (Figure 7). It is open at this point whether this is a mere coincidence or whether it reflects a more general effect. In view of this, it is also referred to as the enhanced value of the latent heat that was reported for the melting of tin clusters (Figure 4).

It should be made clear that the latent heat of melting as displayed in Figure 10 is an average that includes all atoms. In the liquid-shell–solid-core model that

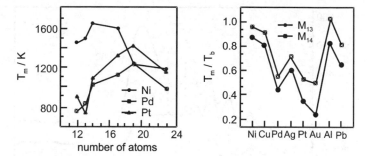

Figure 11 *Melting temperature versus cluster size for nickel, palladium and platinum clusters (left), and overall melting temperature T_m normalised by the experimental bulk melting temperatures T_b for various icosahedral 13 atom and capped icosahedral 14 atom clusters.*
(Reprinted with permission from Ref. 21. Copyright (1996) by the American Physical Society)

applies to region I, the latent heat of melting of the solid core adopts its bulk value. This is plausible since all core atoms have the full coordination number with nearest neighbour atoms, but the fraction of atoms in the liquid layer increases with decreasing cluster size. Thus the *overall* latent heat per atom diminishes.

1.6 Simulated Melting Behaviour of Further Metal Clusters

Size-specific fluctuations of melting temperatures were also predicted for other clusters, as shown in Figure 11 for nickel, palladium and platinum.[21] The trend with size is quite specific for the three elements. Because of the different ranges of the interaction potentials, the melting point is lowered differently for various metals, although the relative melting point of M_{13} is always below that of M_{14}.

1.7 Discrete Periodic Melting of Indium Clusters

The existence of discrete premelting peaks in caloric curves of small clusters has been predicted from Monte Carlo methods.[21] Experimentally, remarkable multiple periodic maxima in nanocalorimetric heat capacity curves were observed for indium nanoclusters with sizes in the range 1000–10,000 atoms (Figure 12) and interpreted as melting peaks of successive completed shells.[22] The analysis is based on the long-known Gibbs–Thomson equation (Equation 15 of Chapter 6) which defines the proportionality of the melting point depression and the inverse cluster radius. For the case of indium, the proportionality constant has been determined experimentally to be 220 ± 10 nm K. The heat capacity measurements can thus be converted to cluster radii. The remarkable finding of Efremov *et al.* is that successive heat capacity maxima correspond to constant increments in the cluster radius between 0.230 and 0.247 nm, which coincide with the lattice parameters of tetragonal indium. The

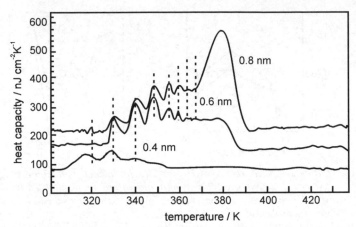

Figure 12   *Heat capacities of 0.4, 0.6 and 0.8 nm average diameter indium clusters on a SiN$_x$ surface giving evidence of discrete periodic melting. Each discontinuity corresponds to the melting of a completed surface monolayer. It should be noted that in the presence of a cluster size distribution the smaller ones melt first.*
(Reprinted with permission from Ref. 22. Copyright (1996) by the American Physical Society)

interpretation is that two factors can contribute to the observation: (i) stable particles corresponding to completed shells require more heat during melting than those of intermediate size and (ii) the number of stable clusters on the support is larger than those of slightly larger or smaller size. While the second interpretation has a well-established basis in other cluster experiments, the first explanation is not supported by the latent heat measurements of sodium clusters (Figure 8), but the reason may lie in the smaller size of the cluster the sodium experiments. Regular size-scaling effects are found only for larger particles, and specific effects are well known to be superimposed for smaller clusters.

Successive intense peaks at increasing temperatures were also observed in differential scanning calorimetry (DSC) traces of suspensions of nanoparticles of the triglyceride trimyristin (Figure 13).[23] This was related to the melting of crystallites with increasing diameter in increments of one unit cell.

1.8   Hydrogen-Induced Melting of Palladium Clusters

Hydrogen is known to react readily with palladium, both at single crystal surfaces and with clusters, and it dissociates and dissolves spontaneously in the bulk. Palladium can therefore be used as a material for hydrogen storage. The intriguing questions now are how hydrogen modifies the metal–metal interaction and whether hydrogen could lower the onset of melting. To investigate this effect, molecular dynamics simulations were performed for the bare icosahedral Pd$_{55}$ cluster and also for Pd$_{55}$ exposed to different amounts of hydrogen.[24] It

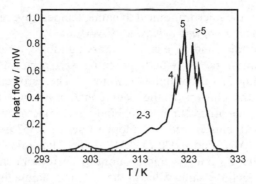

Figure 13 *DSC curve of a trimyristin suspension. The numbers indicate the thickness of the particles, expressed in number of unit cells, melting at the corresponding temperature.*
(Reprinted with permission from Ref. 23. Copyright (1996) American Chemical Society)

was found that hydrogen stabilises the cluster, which is seen in the increase of the binding energy per Pd atom, even though the volume increases as well.

Melting of the bulk system is a first-order transition, characterised by a stepwise increase in the free energy and a singularity in the specific heat. In small systems, the transition is smeared out over a finite temperature interval, resulting in a broadening of the peak in the specific heat. This was seen in the simulation. In addition, the transition shifted substantially to lower temperatures with increasing hydrogen loading. For molecular systems, each chemist is familiar with this effect of cryoscopic freezing point depression, and it is interesting to see that it holds also for a nanoscopic metallic system. Obviously, the softening of the metal–metal bond and the concomitant volume expansion increase the flexibility of the lattice so that the transition to the liquid state becomes easier.

2 Dynamics of Metal Clusters

Traditionally, phase transitions belong to the realm of thermodynamics and statistical physics. Phase transitions of finite size systems of small to moderate size, however, can be understood in dynamical terms. Fundamental and of special relevance to the problems of phase changes in this context are the concepts of temperature, of equipartitioning of kinetic energy and of the degrees of freedom. They were discussed carefully by Jellinek and Goldberg,[25] and a *microcanonical dynamical equipartition postulate* was formulated. It includes a function that plays the role of a finite size correction that becomes unity in the thermodynamic limit. The essence is that in a state of thermal equilibrium, temperature characterises not only the system as a whole but also all parts of it, including the individual atoms and their individual degrees of freedom. Whereas conventionally temperature relates to the canonical ensemble average, the microcanonical dynamic temperature relates to the *time-averaged kinetic energy of any arbitrary subsystem.*

On the basis of this concept, energy conserving molecular dynamics simulations using a Gupta-type many-body potential were performed on aluminium clusters comprising 7, 13, 55 and 147 atoms.[25] The zero-temperature lowest energy form of the clusters is the pentagonal bipyramid for Al_7 and the icosahedron for the other three. The binding energies per atom show the expected trend with cluster size and adopt values of 2.353 eV (Al_7), 2.601 eV (Al_{13}), 2.892 eV (Al_{55}) and 3.020 eV (Al_{147}). The dynamical ensemble vibrational temperature T_Ω and its microcanonical correspondent T_ω (which are expected to approach the same value in the thermodynamic limit) are shown in Figure 14 (left panel) along with the corresponding specific heats $(c_v)_\Omega$ and $(c_v)_\omega$

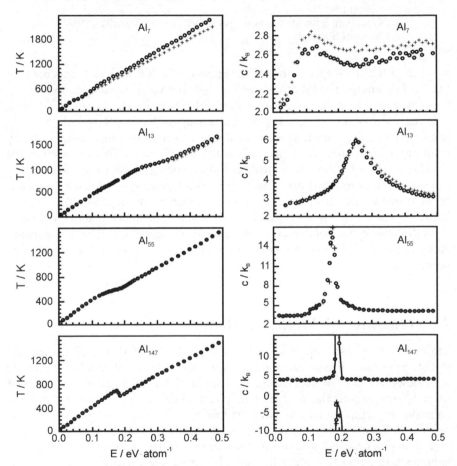

Figure 14   *Dynamical temperatures (left) and specific heats (right) as a function of the internal energy per atom of different size aluminium clusters. The crosses are for the ensemble (Ω-labelled) and the circles for the microcanonical (ω) quantities. (Reprinted with permission from Ref. 25. Copyright (2000) American Institute of Physics)*

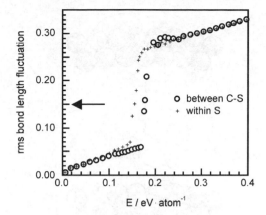

Figure 15 *Root-mean-square (rms) bond-length fluctuations (in units of the equilibrium bond length) as a function of energy for Al$_{13}$ and its two subsystems (C: central atoms, S: surface atoms). The Lindemann criterion places the transition from solid to liquid at a relative rms bond-length fluctuation of 0.10–0.15 (arrow).*
(Reprinted with permission from Ref. 25. Copyright (2000) American Institute of Physics)

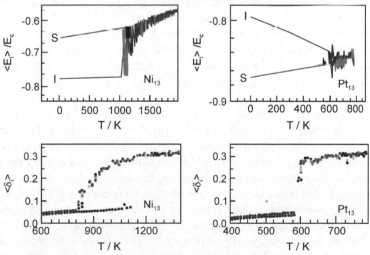

Figure 16 *Monte Carlo simulations of potential energy change (top) and root-mean-square fluctuation (bottom) of each atom as a function of temperature, where E_C is the cohesive energy of the bulk. The labels I and S refer to internal and surface atoms, respectively. A value above 0.1–0.15 of the Lindemann index δ_i indicates a molten state.*
(Reprinted with permission from Ref. 21. Copyright (1996) by the American Physical Society)

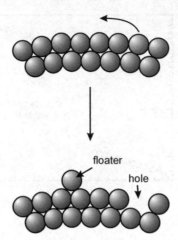

Figure 17   *Promotion of an atom onto the surface of the outer shell of a cluster, leaving behind a hole. Such a floater atom enhances considerably the dynamics in the now incomplete shell*

as a function of the total internal energy per atom. The difference between the two quantities is obvious for the Al_7 at higher energies, but it disappears essentially already for the next larger system. A maximum in the specific heat indicates a solid–liquid phase transition. It is obvious that with increasing system size the height of the maximum in the specific heat increases and the width of the peak decreases. For large systems the peak in c_v is expected to evolve to a point singularity as an indication of bulk melting in a phase transition of first order. Interestingly, for Al_{147}, T does not show a smooth increase with internal energy; rather, there is an intermediate decrease just below an energy of 0.2 eV per atom. This change of slope is reflected in negative values of the specific heat, as discussed above (Figure 9) and observed for Na_{147}, notably the same number of atoms as the simulation in Figure 14.

Of central interest to the study is the separate analysis of the dynamics of the different subsystems. The method is illustrated on the basis of the example of the Al_{13} cluster which is subdivided into two subsystems: the first one is the central atom, and the second one is the remaining twelve atoms at the surface which are equivalent in the low-temperature limiting rigid structure. A suitable indication of the dynamics is given by the root-mean-square bond-length fluctuation between the central atom and the surface atoms on the one hand, and among the surface atoms on the other hand. Figure 15 shows that this fluctuation at first scales linearly with energy. Then there is an abrupt increase in the fluctuation, first between the surface atoms, and only at a somewhat higher energy also between the central atom and the surface shell. This abrupt increase is a signature of a phase-like transition. At first the surface atoms become mobile and can interchange the positions, while the central atom remains at its original site. Only the second step indicates that the central atom can now also exchange with an atom from the outer shell. Also the analysis of the other clusters shows that the different shells form different dynamic classes of dynamical similarity, but after the phase transition has been completed the behaviour of the different shells becomes indistinguishable. This is interpreted as completed melting, and the interesting conclusion is that the surface shell 'melts' at lower energy than the remaining shells. This correlates with the different coordination numbers, which are 12 for the central atom but only 6 for the surface atoms of the Al_{13} icosahedron.

Subshell resolved potential energy changes as a function of temperature and root-mean-square fluctuations were computed by Lee *et al.*[21] The results for 13-atom icosahedral nickel and platinum clusters are shown in Figure 16. For Ni_{13}, the central atom is more stable than the surface atoms at low temperature. Interestingly, this sequence is inverted for Pt_{13}. Since the symmetry of the two clusters is the same, this can only be a consequence of the range of the potential function or the degree of electron delocalisation and band formation. For the same reason, as seen in the root-mean-square fluctuation, Ni_{13} shows surface premelting while central atom and surface shell of Pt_{13} melts at once.

Simulations of Lennard-Jones clusters such as Ar_{147} reveal *surface premelting* as well.[26] Animations show that typically one or a few atoms from the liquid surface shell are promoted so that they float on top of the surface, leaving behind a vacancy (Figure 17). This vacancy enhances significantly the dynamics and diffusion in this outermost layer, and it is accompanied by additional configurational entropy. In more general words, it is the reduced density which permits diffusion and which leads to the impression of a melted surface.

Monte Carlo and molecular dynamics calculations were carried out on Ar_{13} clusters in zeolite L.[27] While the free cluster prefers an icosahedral geometry and melts around 27 K, the equilibrium geometry inside the zeolite pore is

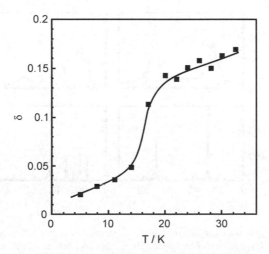

Figure 18 *Variation of the root-mean-square pair distance fluctuation δ of Ar_{13} in zeolite L with temperature*
(Reprinted with permission from Ref. 27. Copyright (1997) American Chemical Society)

different and has an elongated shape that matches the pore geometry. The root-mean-square pair distance fluctuation shows a clear increase between 14 and 20 K which reveals a rigid–non-rigid transition that reminds of the melting transition of the free cluster (Figure 18). It is associated with a sharp increase in cluster volume and with the *inner* atoms becoming mobile, while the outer layer atoms which are in close proximity of the zeolite wall continue to be comparatively immobile. This contrasts with the behaviour of a free cluster of 40 atoms in which the outer shell of atoms acquire mobility before the inner

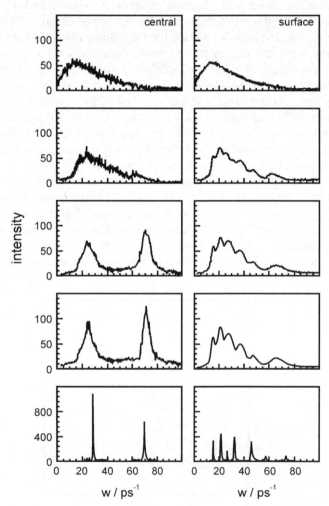

Figure 19 *Power spectra as a function of cyclic frequency w for central and surface atoms of an Al$_{13}$ cluster. The spectra from top to bottom correspond to energies of 0.281, 0.176, 0.166, 0.141 and 0.031 eV per atom, respectively. 63 ps^{-1} correspond to 1000 cm^{-1}*

core. The interaction of the cluster with the zeolite wall thus has the effect of reversing the dynamics from surface melting to *inverse surface melting*.

The cluster dynamics is characterised by the power spectra which are calculated on the basis of the velocity autocorrelation function of each sub-system of an Al_{13} cluster separately (Figure 19). Discrete spectra are predicted for the lowest energy, and the central atom participates essentially only in two modes in which it is slightly coupled to the surface atoms. The narrow lines correspond to nearly undamped atomic oscillations. The surface atoms themselves are involved in several more modes. As the energy increases, the spectrum of the surface atoms first broadens, indicating surface melting, and only at an energy of 0.176 eV per atom the spectral structure of the central atom is washed out as well. A structureless smooth spectrum is obtained for the non-periodic motion of random diffusion in the liquid.

Canonical and microcanonical Monte Carlo simulations of the melting of palladium clusters were carried out by Westergren *et al.*[28] Clusters of 13, 54, 55, 147 and 309 atoms were found to exhibit caloric curves with the typical S-bend at melting and coexistence between solid and molten regions that is expected for finite size systems. In the coexistence region the Pd_{13} cluster constantly switches phase, while only a few switch events were found for Pd_{55} and none for Pd_{147}. In contrast, Pd_{34} was found to melt without an accompanying peak in heat capacity, and at melting the atoms become mobile without any significant change in the geometric structure. Melting of Pd_{34} is therefore a purely entropic phenomenon.

Key Points

- The melting temperature and the latent heat of melting of pore-confined clusters are in general below their bulk values and scale approximately as $1/r$.
- Within the scaling regime, the melting temperature and the latent heat of melting of free clusters are also below their bulk values. However, at small size the behaviour becomes discontinuous because of shell-closing phenomena, and there are cases where the melting temperature can be considerably *above* its bulk value, indicating structural peculiarities with partly stronger bonds.
- By conversion of kinetic energy into potential energy the heat capacity of an isolated ensemble can become negative, *i.e.* the system cools down as more heat is added.
- A regime of a system may be judged molten when the root-mean-square displacement of the atoms exceeds 10–15% of the equilibrium bond length (Lindemann criterion).
- Experimentally, the discreteness of vibrational spectra may serve to distinguish between solid and liquid clusters.
-

The surface of a cluster normally melts before its core. This is related to the lower coordination number of surface atoms. This situation may be reversed in pores.

- Surface melting of completed shells is enabled by promotion of an atom onto the surface of the outer shell (floater atom).

General Reading

- H. Haberland, Melting of clusters, in *Atomic Clusters and Nanoparticles*, C. Guet, P. Hobza, F. Spiegelman and F. David (eds), Springer, Berlin, 2002.
- M. Schmidt and H. Haberland, Phase transitions in clusters, *C. R. Physique*, 2002, **3**, 327.
- J. Jellinek (ed), *Theory of Atomic and Molecular Clusters*, Springer, Berlin, 1999.
- J. Jellinek and A. Goldberg, On the temperature, equipartition, degrees of freedom, and finite size effects: applications to aluminum clusters, *J. Chem. Phys.*, 2000, **113**, 2570.

References

1. H. Haberland, *Melting of clusters*, in *Atomic Clusters and Nanoparticles*, C. Guet, P. Hobza, F. Spiegelman and F. David (eds), Springer, Berlin, 2002.
2. Ph. Buffat and J.-P. Borel, *Phys. Rev. A*, 1976, **13**, 2287.
3. K. Koga, T. Ikeshoji and K. Sugawara, *Phys. Rev. Lett.*, 2004, **92**, 115507.
4. S.L. Lai, J.Y. Guo, V. Petrova, G. Ramanath and L.H. Allen, *Phys. Rev. Lett.*, 1996, **77**, 99.
5. T. Bachels, H.-J. Güntherodt and R. Schäfer, *Phys. Rev. Lett.*, 2000, **85**, 1250.
6. R. Kofman, P. Cheyssac and F. Celestini, *Phys. Rev. Lett.*, 2001, **86**, 1388.
7. A.A. Shvartsburg and M.F. Jarrold, *Phys. Rev. Lett.*, 2000, **85**, 2530.
8. G.A. Breaux, R.C. Benirschke, T. Sugai, B.S. Kinnear and M.F. Jarrold, *Phys. Rev. Lett.*, 2003, **91**, 215508.
9. S. Chacko, K. Joshi, D.G. Kanhere and S.A. Bludell, *Phys. Rev. Lett.*, 2004, **92**, 135506.
10. A.N. Goldstein, C.M. Echer and A.P. Alivisatos, *Science*, 1992, **256**, 1425.
11. A.P. Alivisatos, *J. Phys. Chem.*, 1996, **100**, 13226.
12. M. Schmidt, R. Kusche, W. Kronmüller, B. von Issendorff and H. Haberland, *Phys. Rev. Lett.*, 1997, **79**, 99.
13. M. Schmidt, R. Kusche, B. von Issendorff and H. Haberland, *Nature*, 1998, **393**, 238.

14. M. Schmidt and H. Haberland, *C. R. Physique*, 2002, **3**, 327.
15. R. Kusche, Th. Hippler, M. Schmidt, B. von Issendorff and H. Haberland, *Eur. Phys. J. D*, 1999, **9**, 1.
16. H. Haberland, T. Hippler, J. Donges, O. Kostko, M. Schmidt and B. von Issendorff, *Phys. Rev. Lett.*, 2005, **94**, 035701.
17. D. Lynden-Bell, *Physica (Amsterdam)*, 1999, **263A**, 293.
18. P. Labastie and R.L. Whetten, *Phys. Rev. Lett.*, 1990, **65**, 1567.
19. M. Schmidt, R. Kusche, T. Hippler, J. Donges, W. Kronmüller, B. von Issendorff and H. Haberland, *Phys. Rev. Lett.*, 2001, **86**, 1191.
20. S.J. Zhao, S.Q. Wang, D.Y. Cheng and H.Q. Ye, *J. Phys. Chem. B*, 2001, **105**, 12857.
21. Y.J. Lee, E.-K. Lee, S. Kim and R.M. Nieminen, *Phys. Rev. Lett.*, 2001, **86**, 999.
22. M. Yu Efremov, F. Schiettekatte, M. Zhang, E.A. Olson, A.T. Kwan, R.S. Berry and L.H. Allen, *Phys. Rev. Lett.*, 2000, **85**, 3560.
23. T. Unruh, H. Bunjes, K. Westesen and M.H.J. Koch, *J. Phys. Chem. B*, 1999, **103**, 10373.
24. H. Grönbeck, D. Tománek, S.G. Kim and A. Rosen, *Z. Phys. D*, 1997, **40**, 469.
25. J. Jellinek and A. Goldberg, *J. Chem. Phys.*, 2000, **113**, 2570.
26. R.S. Berry and H.-P. Cheng, *Phase changes for clusters and for bulk matter*, In: *Physics and Chemistry of Finite Systems From Clusters to Crystals*, **Vol I**, P. Jena, S.N. Khanna, B.K. Rao, (eds.), Kluwer Academic, Dordrecht, 1992.
27. R. Chitra and S. Yashonath, *J. Phys. Chem. B*, 1997, **101**, 389.
28. J. Westergren, S. Nordholm and A. Rosén, *Phys. Chem. Chem. Phys.*, 2002, **5**, 136.

CHAPTER 9

Phase Transitions of Two-Dimensional Systems

1 Melting of Thin Layers

The surface-to-volume ratio of two-dimensional systems scales with the inverse thickness so that behaviour analogous to that of three-dimensional systems should be expected. This was tested for chalcogenide alloys of the composition $GeSb_2Te_4$, and the results arc displayed in Figure 1. The fitted curve corresponds to a melting temperature of $(786\ K{-}354d^{-0.7})$ for the thin plates (left entry) and of $(786\ K{-}558r^{-0.75})$ for the spherical clusters. The exponent of $d(r)$ does not correspond to -1.0 as expected on the basis of the Gibbs–Thomson (Equation 15 of Chapter 6). The deviation would have to be accommodated in a liquid surface layer of a certain thickness t_0 (Equation 20 of Chapter 7).

2 Structural Phase Transitions of Thin Layers

In an analogous way as spherical clusters, nanolayered materials often exhibit structural features, which are significantly different from those of their bulk counterparts.[2] A series of structural transitions was observed in sputter deposited Ti/Al multilayered thin films, employing equal thickness of the two components with repeat periodicities in the range of 5–100 nm. The structure of the Ti layer changed from hexagonal close-packed (hcp) to face-centered cubic (fcc) on reducing the layer thickness below 5 nm and reverted back to an hcp structure below a layer thickness of 2.5 nm. The Al layer changed from fcc to hcp below a layer thickness of 2.5 nm.[2] An explanation of this interesting phenomenon involves a combination of the changes in volume free energy (ΔG_V), strain free energy (ΔG_S) and interfacial energy (γ), during a phase transition:

$$\Delta G = \Delta G_V V + \Delta G_S V + \gamma A \tag{1}$$

where V represents the volume and A the area. After division by the area the total free energy change of a multilayer consisting of Ti and Al with fractions f_{Ti} and f_{Al} respectively, and with a periodicity $\lambda = d_{Ti} + d_{Al}$ can be written as

$$\Delta G/A = \Delta G_{Ti}f_{Ti}\lambda + \Delta G_{Al}f_{Al}\lambda + 2\gamma \tag{2}$$

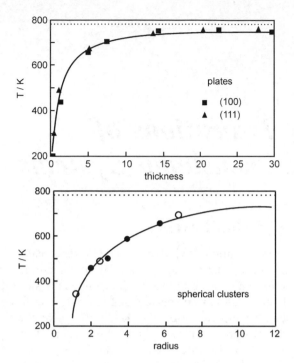

Figure 1 *Melting point versus plate thickness of GeSb₂Te₄ chalcogenide thin films (upper panel) and spherical clusters (lower panel). Thickness and radius are given in units of the lattice constant. The broken lines indicate the bulk melting point. (Reproduced with permission from Ref. 1.Copyright (1997) TMS)*

where ΔG_{Ti} and ΔG_{Al} now contains both, volume and strain free energy. The result is shown schematically in Figure 2 for a realistic hierarchy of thermodynamic values. It can explain the experimental findings, and it illustrates again the competition between bulk and interfacial contributions to the total free energy and suggests that thin film strain energy plays a pivotal role in determining the structures.

Other interesting examples which reflect the competition between volume energy and interfacial strain have been studied by electrochemical deposition of copper on gold and silver single crystal surfaces.[3,4] Cu, Ag and Au all have an fcc equilibrium structure, and the lattice constants of Au and Ag are close to the same, while that of Cu is significantly smaller. Cu deposition onto Ag starts with the formation of a complete monolayer. This process takes place in a potential region that is positive with respect to the Nernst potential for bulk Cu deposition. This so-called *under-potential deposition* generally occurs when the deposit–substrate interaction (here Cu–Au) exceeds that between the atoms of the deposit (Cu–Cu). In the case of Au(111) and Au(100) as a substrate, a pseudomorphic Cu monolayer is formed at under-potentials, that is the Cu atoms occupy substrate lattice sites. This results in a strained Cu monolayer because the next neighbour distance for Au is about 13% larger than for Cu.[4]

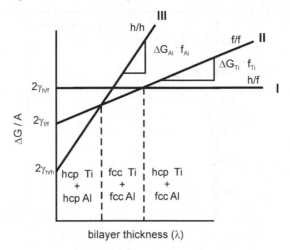

Figure 2 *Schematic representation of the basic thermodynamic model proposed for explaining the structural transitions in Ti/Al multilayers. Each of the three lines represents the free energy change involved in the possible transitions with respect to the bulk stable structures hcp Ti and fcc Al. In the range where line I is lowest the structures correspond to those of the bulk, in the intermediate range of line II Ti changes phase, and in range III Al adopts a different phase.* (Reproduced with permission from Ref. 2. Copyright (1997) TMS)

No under-potential deposition of Cu was observed on Ag(111), but again pseudomorphic growth was found. This is in agreement with the general observation that only the very first or the first two layers are deposited pseudomorphic, if at all, limited by the stress due to the crystallographic misfit with the substrate. However, for Cu on Ag(100) the first eight(!) layers grow pseudomorphic. This was explained with the initial growth of body-centered cubic (bcc) Cu on Ag(100), which results in a vanishingly small lattice mismatch with the substrate. Obviously, this can offset the energy difference of about 20 meV by which bcc Cu is thermodynamically less stable than fcc Cu. With the 9th Cu layer, however, a sudden structural transition occurs in the overlayer which leads to a buckled surface and, as deposition continues, to fcc Cu.[3]

A similar observation was made for the deposition of Cu on Au(100), with the main difference that the first 10 layers grow pseudomorphic as bcc Cu before the structure flips to fcc on deposition of the 11th layer.[5] Reduction of the next-neighbour distance allows the accommodation of 11 Cu atoms on top of 10 Cu atoms of a pseudomorphic layer, which leads to the observed stripes reflecting modulated buckling.

3 Glass Transition of a Polymer Thin Film

The phase transition behaviour of free-standing thin films is quite spectacular but qualitatively analogous to the melting point depression of spherical

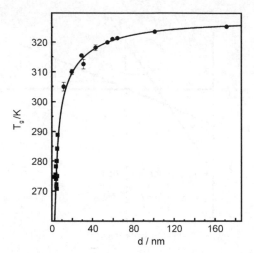

Figure 3   *Glass transition temperature in polystyrene thin film. The line represents a single parameter fit to the data of an expression derived from mode coupling theory*[6]

clusters. The glass transition temperature of polystyrene films drops by more than 50 K when the film thickness reduces to *ca.* 5 nm (Figure 3).

4   Surface Alloy Phases

A number of different phenomena are observed when one metal is deposited onto another one.[7] The deposited metal may form islands or it may alloy into the first or the first several layers. Alloying may take place both in cases where the two metals form an alloy in the bulk, and more interestingly, in cases where they do not alloy in the bulk. For example, Ag and Ni show absolutely no tendency for alloying in the bulk. However, Au can replace Ni in the first surface layer under the formation of a surface alloy.[8] The effect has been studied in detail using density functional theory, and a database has been constructed of segregation energies and surface mixing energies for all combinations of transition metals.[7] On this basis it is possible to derive phase diagrams for surface alloys.

Key Points

- The phase transition behaviour of thin layers follows much the same scaling behaviour as those of clusters.
- The structure of thin layers is determined by a competition between volume energy and interfacial energy, in particular strain due to lattice mismatch.
- Metals which do not form alloys in the bulk may form surface alloys in the first or the first several layers.

References

1. J.K. Lee, B.K. Cheong, W.M. Kim and S.G. Kim, *A simulation study on the melting of nanocrystalline plates and spherical clusters*, in *Chemistry and Physics of Nanostructures*, E. Ma, B. Fultz, R. Shull, J. Morral and P. Nash (eds), The Minerals, Metals & Materials Society, Warrendale, PA, 1997.
2. R. Banerjee, X.D. Zhang, R. Ahuja, M. Asta, A.A. Quong, S. Dregia and H.L. Fraser, *Structural transitions in Ti/Al nanolayered thin films*, in *Chemistry and Physics of Nanostructures*, E. Ma, B. Fultz, R. Shull, J. Morral and P. Nash (eds), The Minerals, Metals & Materials Society, Warrendale PA, 1997.
3. M. Dietterle, T. Will and D.M. Kolb, *Surf. Sci.*, 1998, **396**, 189.
4. R. Randler, M. Dietterle and D.M. Kolb, *Z. Phys. Chem.*, 1999, **208**, 43.
5. R.J. Randler, D.M. Kolb, B.M. Ocko and I.K. Robinson, *Surf. Sci.*, 2000, **447**, 187.
6. S. Herminghaus, R. Seemann, D. Podzimek and K. Jacobs, *Nachrichten aus der Chemie*, 2001, **49**, 1398.
7. A. Christensen, A.V. Ruban, P. Stoltze, K.W. Jacobsen, H.L. Scriver, J.K. Nørskov and F. Besenbacher, *Phys. Rev. B*, 1997, **56**, 5822.
8. L. Pleth Nielsen, F. Besenbacher, I. Stensgaard, E. Lægsgaard, C. Engdahl, P. Stoltze, K.W. Jacobsen and J.K. Nørskov, *Phys. Rev. Lett.*, 1993, **71**, 754.

CHAPTER 10
Catalysis by Metallic Nanoparticles

1 Some General Principles of Catalysis by Nanoparticles

Most metal catalysts of practical relevance are nanoparticles rather than large single crystal surfaces. What is it that clusters of atoms with often irregular shape can do chemistry more easily than single crystal surfaces? There are several reasons for this.

First of all, catalytic activity often takes place at defects on surfaces, and on steps and kinks. They allow for special bonding situations which are favourable for the breaking and making of bonds. The irregular surface of nanoparticles offers a suitable environment with a rich variety of such defects. Second, surfaces restructure continuously to adapt to the adsorbates and thus to the chemistry that occurs. Atoms near the surface of small clusters are much more flexible to move since there are fewer neighbours than in large single crystals, and thus restructuring is much easier. Often, surface restructuring is adsorbate-induced. A third reason is based on the electronic properties of a cluster. They relate to the availability of electrons for redox processes and for the formation of chemical bonds to adsorbate molecules. It is the level of the Fermi energy, ε_F (in more chemical terms: the energy of the *frontier orbitals*), and the width of the valence d-band of transition metals which determine the electronic properties of relevance to chemistry, and there are several ways of influencing it.

- For small free clusters, the Fermi energy depends on the cluster size. This is addressed in Chapter 4.
- Tensile strain increases the interatomic distance and influences the band structure. It has been observed that chemisorption energy and reactivity can be much different on strained regions of a surface.
- An isolated guest atom in a metal distorts the Fermi surface in its local environment.
- Bimetallic alloys with similar mole fractions have a Fermi surface at an equilibrated, intermediate value between those of particles of the individual plain metals that have the same size.

- For supported clusters, the metal–support interaction can cause appreciable changes in ε_F via charge polarisation.
- If the catalyst particles rest on a conducting support (an electrode), the Fermi level can also be tuned by application of an external bias voltage.

The particles of real metal catalysts are normally of irregular shape, and they often exhibit a broad size distribution. Moreover, they are normally supported by a material which may itself not be of a simple homogeneous nature, and they may be alloyed or in contact with electropositive promotor ions, such as K^+ or Cs^+. Furthermore, they may be partly oxidised or 'poisoned' by trace impurities which accumulate on a surface due to their high chemisorption energy. In view of this complexity it is plausible that it has been notoriously difficult to generate reproducible catalytic materials and to disentangle the influence of individual variables in such a multi-parameter system. Surface science has taken the approach to study elementary reactions on well-defined single crystal surfaces. They are only crude models of real catalysts, but they allow discerning some important principles. In the following we will focus on a number of studies, which are suitable to illustrate the tunability of electronic properties of catalytic nanomaterials.

Chemists are used to think in terms of the one-electron approximation in molecular orbital (MO) theory. This is a useful zero-order approach also for the discussion of the interaction between an adsorbate and a catalyst metal. All transition metals in the metallic state have a broad, half-filled s-band that is centred at the Fermi level. Interactions of this band with an adsorbate atomic orbital result in a broad state which is mostly below the Fermi energy and thus provides only weak chemisorption bonding. d-bands are much narrower and, therefore, give rise to discrete and clearly split bonding and antibonding states. If the energy of the antibonding state is above the Fermi level so that it is an empty lowest unoccupied molecular orbital (LUMO), this leads to a strong chemisorption bonding. The situation is illustrated in Figure 1.

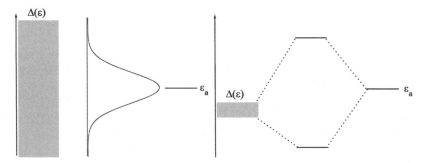

Figure 1   *The local density of states of an adsorbate orbital interacting with a broad surface band (a) and with a narrow band (b). Case (a) corresponds to the interaction with a metal s-band and leads to weak chemisorption; case (b) is representative for the interaction with a transition metal d-band*
(Reproduced with permission from Ref. 1)

2 Size-Controlled Catalytic Clusters

Small unsupported (gas phase) transition metal clusters exhibit strong variations as a function of size not only in their physical and electronic properties but also in their chemical behaviour. With the exception of coinage metals, orders of magnitude changes in the rate constant of dissociative hydrogen chemisorption near room temperature have been reported for many transition metal clusters.[2] Consistent with bulk behaviour, H_2 reacts readily with most transition metal clusters larger than about 30–40 atoms, while small alkanes such as methane and ethane react preferentially with very small clusters of specific metals with high ionisation potential such as platinum and palladium.

Obviously, the ability to accept or donate charge plays a key role. H_2 bond activation takes place on bulk metal surfaces with low work function and on large clusters with low ionisation potential which act as charge donor to the adsorbate molecule. But it also works well on small clusters with a sufficiently high electron affinity so that the metal can act as charge acceptor.[2] The much different behaviour of small alkanes has been ascribed to their much poorer electron accepting potential compared to that of hydrogen due to the high energy of their σ^* antibonding orbital, while they are considered equally good charge donors.

The higher activation barriers in reactions of coinage metals towards H_2 and O_2 is generally attributed to their filled d-bands ($d^{10}s^1$ configuration), which gives them extra stability. Bulk gold is thus commonly found to be chemically extremely inert and inactive as a catalyst.[3] It was, therefore, a surprise when highly dispersed gold particles were found to exhibit high catalytic activity.[4] A study of the reactivity of small gold clusters consisting of less than 15 atoms revealed a pronounced dependence on cluster size and cluster charge.[2] For example, cationic clusters react readily with D_2, while anionic clusters appear to be inactive under the same conditions. Similar observations were made for methane. With oxygen, only even atom (odd electron) anionic clusters are reactive, and Au_{10}^+ seems to be the only cation with some reactivity. Also the total uptake of deuterium or methane on gold clusters depends on size, with D/Au varying from a factor of 3 for the gold dimer to zero for clusters containing more than 15 gold atoms.[2]

Heiz et al.[5] demonstrated that the dissociation of CO on size-selected supported nickel clusters is strongly dependent on the exact cluster size as previously shown for free clusters by Vajda et al.[6]

Baltenau et al.[7] measured the gas phase reactivity of size-selected anionic and cationic platinum clusters with N_2O. In the reaction, an oxygen atom is added to the cluster, accompanied with the loss of N_2. The results which are displayed in Figure 2 demonstrate that the reactivity varies by nearly three orders of magnitude in a seemingly random manner. Obviously, each atom counts. Of the cationic clusters, Pt_n^+, $n = 6$–9, 11, 12, 15 and 20 are very reactive, while $n = 10$, 13, 14, 19 and 21–24 show very little to no reactivity. For the anionic clusters, the variability is somewhat lower and the size dependence is considerably different. While the $n = 6$ cluster is the fastest among the cations, it is the

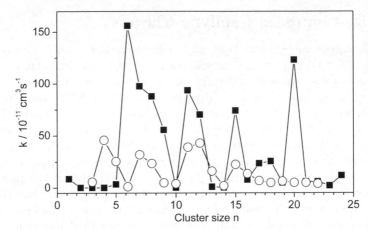

Figure 2   *Absolute rate constants for the reaction with N_2O of cationic (squares) and anionic (open circles) Pt_n clusters. Some of the lowest values represent upper limits of the rate constant for unreactive cluster sizes*
(Drawn using data from Ref. 7)

slowest among the anions. On the other hand, $n = 10$ and 14 represent the minima in both charge states. This behaviour is not yet understood. It is therefore also not possible to interpolate for the behaviour of neutral clusters.

Titania (TiO_2)-supported Au particles have been found to be active at ambient conditions for the oxidation of CO to CO_2. The activity of gold particles is very sensitive to their size. Figure 3 shows that particles in the range of 2–3 nm are most active.[8] Enhanced activity was ascribed to the larger adsorption energy of O and CO on steps. If the first layer on the support is not counted, since the Au atoms in this layer are responsible to bonding of the cluster to the support, then a reactivity maximum is obtained for double layer islands, which is in reasonable agreement with experimental observations.[9] The effect demonstrates the possibility of tuning the electronic properties of the metal particles to achieve high catalytic activity. Since titania is an important ingredient of white paint the effect can be used to reduce the levels of CO pollution in buildings. Furthermore, 3–5 nm gold particles have been developed commercially by a Japanese company as bathroom 'odour eaters'.

The reactivity of gold nanoparticles of 1–8 nm diameters was investigated also on a silicon wafer support, isolated by a thin layer of native SiO_2.[10] The clusters were treated first in hydrogen plasma to ensure that they were elemental gold without any detectable traces of oxygen. Subsequently, they were exposed to oxygen atoms and radicals in oxygen plasma and then analysed by X-ray photoelectron spectroscopy (XPS). Figure 3 shows the two elemental gold peaks, marked by vertical lines, and the corresponding lines in the oxide which are shifted to higher binding energy by ca. 2 eV. The spectrum of the 1.4-nm cluster, corresponding to Au_{55}, differs dramatically from the others. It shows only the elemental gold signals and demonstrates that gold-55 is much more oxidation resistant, suggesting that its properties as a catalyst are also different from those

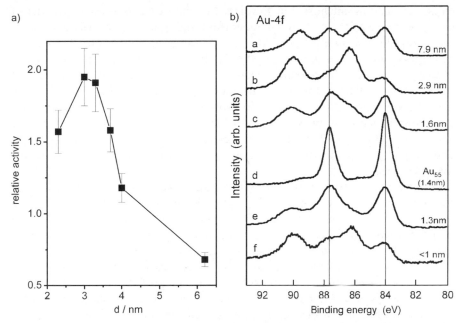

Figure 3   *(a) Effect of particle size on the activity of titania-supported gold particles for the oxidation of CO.[8] (b) Au-4f photoelectron spectra obtained after exposure of pure gold particles of 1–8 nm diameters to reactive oxygen provided by rf plasma. The vertical lines indicate the positions corresponding to elemental gold; the other two peaks belong to the oxide*
(Reprinted with permission from Ref. 10. Copyright (2002) AAAS)

of smaller or larger clusters. We recall that gold-55 was also found to be exceptional as regards structure, as it was found to be amorphous, whereas clusters of this size of neighbouring elements have a well-defined icosahedral structure (see Chapter 3, Section 2).[11]

Also supported nanosize particles of gold on various oxides as well as atomic double layer gold islands of up to 4 nm diameter on titania exhibit enhanced catalytic activity, in particular for the low temperature oxidation of carbon monoxide (CO).[12] While this may turn out to be of great significance to environment friendly sensor and chemical technologies, the reasons for the unexpected behaviour are not yet understood. From the normal size dependence of the metal work function, which increases towards the atomic ionisation potential (see Figures 12 and 15 of Chapter 4), it is clear that small neutral clusters should not be expected to help oxidation by temporary electron transfer better than bulk metal. In view of this, it is significant that a gold octamer (Au_8) is adsorbed rather strongly (by 5.56 eV) on MgO(100) surfaces containing oxygen vacancy F-centres, and that this adsorption is accompanied by charge transfer of ca. 0.5 e into the cluster. Similar partial charging of the Au_8 cluster occurs also on a defect-free MgO(100) surface, but the binding energy to the surface is reduced to about 40% of its value near an F-centre. The enhanced binding anchors and localises the charged clusters near the defect sites and prevents them from agglomeration and concomitant deactivation of the catalyst. It is interesting that early investigations of molecular oxygen adsorption on free clusters occurred only for even numbered anionic clusters (Au_{2n}^-, $2 \leq n \leq 10$), and the reactivity correlated with the size-dependent pattern of the electron affinities of the neutral gold clusters (Ref. 18 of Ref. 12). This demonstrates that the charging is essential, and it suggests that the neutral gold clusters deposited on MgO act as a mediator of electron transfer from the surface to the adsorbed substrate. The electronic interaction between catalyst and support is a key feature which has been known to exist for some time for many catalysts.

An additional aspect emerges from the fact that there is a minimum cluster size for the oxidation of CO on MgO-supported Au_n. Clusters below $n = 8$ are inactive. For $n \geq 8$ the catalytic activity oscillates to some extent, as shown in the pioneering work by Sanchez et al.[12] (Figure 4). It was found that in the absence of oxygen, CO desorbs from Au_8 at temperatures between 150 and 180 K. Nevertheless, in presence of O_2, catalytic activity was observed also above the CO desorption temperature. This was explained and supported in extensive ab initio simulations by two different reaction mechanisms. O_2 was shown to readily adsorb at several sites in all cases, forming an activated but not dissociated peroxo molecular state with a weakened, highly stretched bond length (0.141–0.146 nm) compared to that of the free molecule (0.124 nm). The formation of this state involved the occupation of the antibonding π^* oxygen molecular orbital, benefiting from the partially negatively charged adsorbed gold cluster. In the low temperature mechanism, O_2 and CO are initially adsorbed, both on the top facet of Au_8. This permits reaction to weakly adsorbed CO_2 and atomic oxygen via a rather low energy barrier of 0.1 eV. The second reaction channel, which benefits

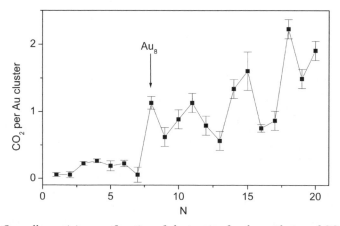

Figure 4   *Overall reactivity as a function of cluster size for the oxidation of CO to CO_2 on gold clusters supported on defect rich MgO(100).*
(Reprinted with permission from Ref. 12. Copyright 1999) American Chemical Society)

particularly from the defect-rich support, occurs via direct reaction of gaseous CO with preadsorbed O_2, most remarkably without any energy barrier.

A very clear and obviously nonlinear dependence on cluster size was also found for CO oxidation on size-selected Pt_n clusters ($5 \leq n \leq 20$) on the Mg(100) films (Figure 5).[13] The reactivity per cluster increases abruptly in going from Pt_8 to Pt_{15}. The reactivity per atom has a maximum for Pt_{15} and a minimum for Pt_8; in addition, there is a local minimum for Pt_{13}, which corresponds to a magic cluster size with the first atomic shell closing for both icosahedral and cuboctahedral structure.

Dissociation of O_2 occurs when there is sufficient back donation of electron density from the clusters into the antibonding π_g^* state or enough donation from the π_u or σ_g orbital into the cluster. Energetic degeneracy of the cluster orbital with any of these oxygen molecular orbitals maximizes the interaction and facilitates the transfer of electron density. The level diagram displayed in Figure 6 suggests that the maximum reactivity of Pt_{15} could indeed be ascribed to a maximum back donation due to the degeneracy of the oxygen π_g^* state with the highest occupied molecular orbital (HOMO) of Pt_{15}.

It should be noted that moisture in the ppm range plays a very significant role in accelerating CO oxidation on supported gold nanoparticles without changing the activation energy.[14]

It is often believed that decrease of specific catalytic surface area is the only consequence of particle agglomeration. Considering that quantum size effects influence the levels of HOMO and LUMO orbitals, it has to be expected that electronic effects should show up. Indeed, it has recently been demonstrated

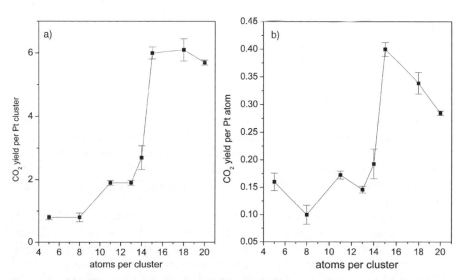

Figure 5 *Total number of catalytically produced CO_2 molecules per Pt cluster (a) and per Pt atom in a cluster (b) as a function of cluster size.*
(Reprinted with permission from Ref. 13. Copyright (1999) American Chemical Society)

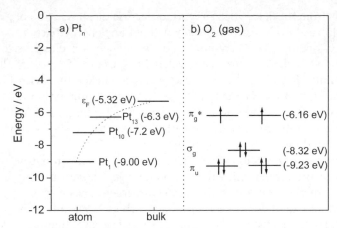

Figure 6 *Energy diagrams of the relevant electronic states of Pt_n including the individual atom and the bulk limit, and of molecular oxygen. Dissociation of O_2 is favoured when the energy of the cluster HOMO state is close to that of π_g^* of O_2. (Reprinted with permission form Ref. 13. Copyright (1999) American Chemical Society)*

that peak potentials in cyclic voltammetry for CO oxidation on carbon-supported nanoparticles depend on the average size of the metal particles.[15] The peak potential for surface oxide reduction exhibits systematic shifts from the position characteristic of bulk polycrystalline Pt foil towards lower potential as the particle diameter decreases below an average value of ca. 4 nm (Figure 7). This trend is related to stronger bonding of oxygen onto nanoparticles as compared to extended surfaces. On the other hand, the CO stripping peak shifts to higher potentials as the particle size decreases, indicating an increase of the oxidation overpotential.

3   Shape-Dependent Catalytic Activity

In high vacuum surface catalysis studies with single crystals, a great deal of work has been carried out in which crystals cut with different facets are shown to catalyse different reactions. Nanoparticles of different shapes have different facets. Moreover, they have different fractions of atoms located at different corners and edges. Thus, one would expect the catalytic activity to be different in catalysing the same reaction.

An investigation of the electron transfer reaction between hexacyanoferrat(III) and thiosulfate ions, catalysed by platinum particles of different shape, confirmed this view.[16] Tetrahedrally dominated particles were a factor of 2.8 more active than near spherical particles with (111) and (100) facets and close to the same size, and slightly larger dominantly cubic particles were even less reactive (Figure 8). There is also a clear difference in activation energies.

The different reactivities were correlated with the fraction of corner and edge atoms. Moreover, it was concluded that the corner atoms of the tetrahedral

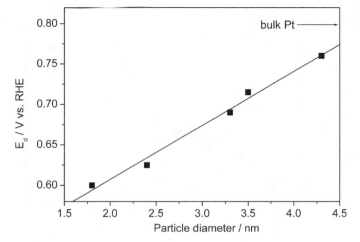

Figure 7 *Potential of the surface oxide reduction peak E_d of glassy carbon-supported Pt nanoparticles as a function of particle size*
(Prepared using data from Ref. 15)

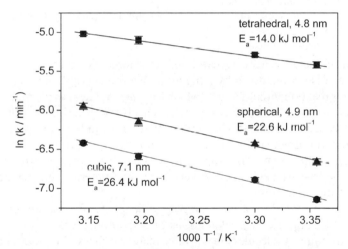

Figure 8 *Shape-dependent catalytic activity of Pt nanoparticles of different shapes and diameters between 4.8 and 7.1 nm in electron transfer between hexacyanoferrat(III) and thiosulfate ions. The preexponential factors increase in the same order as the activation energies, giving evidence of a compensation effect as described by the Meyer–Neldel rule.*
(Redrawn from Ref. 16)

particles are much more reactive, which makes good physical sense since they are more chemically unsaturated with a lower coordination number.

A clear *compensation effect*, also called *Meyer–Neldel rule*, was observed between the apparent Arrhenius preexponential factors and the activation energies. This effect, which is often observed in homogeneously and in

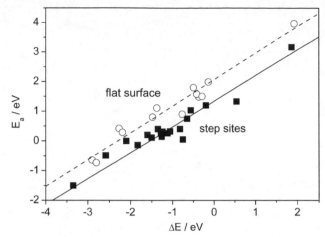

Figure 9 *Calculated activation energies for N$_2$, CO, NO and O$_2$ dissociation on a number of different metals plotted as a function of the calculated dissociative chemisorption potential energy. Open circles are for fcc(111), hcp(0001) and bcc(110) close-packed surfaces, solid squares are for steps (which include fivefold coordinated sites).*
(Redrawn using values from Ref. 17 Copyright (2002) with permission from Elsevier)

particular in heterogeneously catalysed reactions, expresses the fact that a reaction that is slowed down by a higher activation energy is accelerated by an increased preexponential factor (which in turn relates to a more positive apparent activation entropy). It normally reflects the situation that the reaction does not occur in a simple single-step but that a preequilibrium operates before the rate-determining reaction. In catalysis, this preequilibrium may often be an adsorption equilibrium. The apparent Arrhenius parameters are thus composed of the true Arrhenius parameters of the rate determining step and of the preequilibrium constant.

The above correlation of the reactivity with the number of corner and edge atoms is also supported by calculations of activation energies for the dissociation of diatomic molecules on various metal surfaces. In plots against the dissociation energies, it was found that the activation energy is lower in the average by 0.7 eV for step sites than for flat surface sites (Figure 9).[17] This finding is quite independent of the type of molecule and metal because the transition state geometry is more or less the same in all cases.

4   The Effect of Strain

When a material undergoes tensile or compressive strain, this decreases or increases the overlap of orbitals of neighbouring atoms. In metals this leads to a decrease or increase of the bandwidths (see Figure 1 of Chapter 4). When a thin metal slab is bent so that one side is under tensile and the other one under

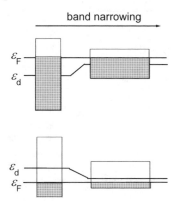

Figure 10 *Effect of band narrowing on the d-band energy, represented by the d-band centre ε_d, relative to the Fermi energy ε_F. Typical transition metal catalysts have more than half-filled d-bands (upper scheme) so that band narrowing correlates with an increase in energy to maintain a constant degree of filling. (Redrawn with permission from Ref. 1)*

compressive strain, this does not lead to charge polarisation. Rather, the degree of filling of the bands near the Fermi level is maintained on both sides. As illustrated in Figure 10, this can happen only when the centre of the d-band, ε_d, of late transition metals is pushed up under tension and that of early transition metals moves down relative to the Fermi level ε_F of the unstrained metal. In the absence of charging effects, this shift of ε_d is often sufficient for an understanding of changes in chemical behaviour.[1]

There are different ways to experimentally realise well-defined strained surfaces. It is possible to epitaxially grow several layers of one metal on top of another, for example by underpotential deposition (see Chapter 9, Section 2) or by vapour deposition. Isomorphic growth results in strained overlayers when the lattice constants of the two metals differ. It has been shown experimentally that the chemical properties of such strained overlayers can be significantly different from those of the unstrained overlayer metal.[18,19] Gsell et al.[20] used an alternative approach. They implanted noble gas ions into a single crystal ruthenium surface. This causes regions of the surface above the implanted atom to buckle out. The middle of this buckled region is under tensile strain, whereas the periphery is subject to compressive strain. It was shown by scanning tunnelling microscopy on a Ru(0001) surface that oxygen atoms adsorb preferentially on sites in the expanded regions of the surface.

Density functional calculations confirm this picture, as shown in Figure 11. The chemisorption energy of both, oxygen atoms and CO molecules, gets stronger in the expanded area and weaker in the compressed region. The chemisorption energy increases by about 0.6% (1.2 kJ mol^{-1}) for CO and by 1.3% (6.7 kJ mol^{-1}) for O per 1% lattice expansion.[21] Simultaneously, the calculated dissociation energy of CO decreases by as much as 14% (11 kJ mol^{-1}) for each 1% of tensile strain, making it much more reactive on the expanded surface.

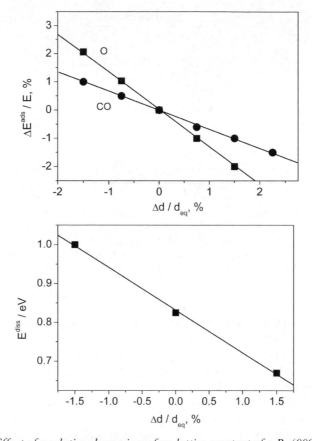

Figure 11 *Effect of a relative change in surface lattice constant of a Ru(0001) surface on the binding energy of atomic oxygen (−5.32 eV on the unstrained surface) and of molecular CO (−2.01 eV on the unstrained surface) in the upper panel, and on the CO dissociation energy in the lower panel.*
(Reprinted using data in Ref. 21. Copyright (1998) by the American Physical Society)

On the basis of a large set of data from similar calculations on other systems it was suggested that this trend is quite general, and that it can be reduced mostly to a single parameter, which is the energy of the d-band centre relative to the Fermi energy as the lattice is strained.[1] The higher the energy of the d-bands, the more likely is it that the orbitals characterising the adsorption bond are split into a pair of bonding and antibonding states, with the antibonding ones located above the Fermi level so that the metal-adsorbate interaction becomes net attractive. The identity of the metal shows up only in the strength of the effect. Following the relative sizes of the d-orbitals, it increases as 3d < 4d < 5d. Furthermore, while the reported experiments and calculations apply to distorted single crystal surfaces, the effects should on the same basis clearly be present for small free and supported clusters, which are known to be severely strained systems. The situation in clusters is more complex, however, so that it is more difficult to separate the strain effect from those of size and geometry and from the interaction between the electrons in the overlayer with those of the support material.

It is interesting in this context to extend the picture of width and position of the d-band (Figure 10) from the general case to that of a local property near surfaces. We have learnt from Figure 6 of Chapter 4 that the total width of a band grows with increasing coordination number and it is plausible that the character of an adatom on a surface is closer to that of a free atom than to that of an atom in the bulk. This means that the bandwidth becomes a local property near a surface. It reduces in the order of the coordination number: bulk > flat surface facets > step > kink > adatom. Thus, to maintain the degree of filling of the band and to avoid charge polarisation, the d-band centre of late transition metals is pushed up more and more towards the Fermi level as the coordination number decreases. In view of what has been said above, it means that adsorption is stronger on steps than on facets. There is indeed ample experimental evidence and theoretical confirmation for this, as demonstrated in Figure 12.[1] In the case of N_2 dissociation on Ru(0001), the reactivity on steps is so much greater than that on terraces that as few as about 1% of step sites on a single crystal surface completely dominate the measured reactivity.

An illustrative example is given in Figure 13 which shows the activation energy for the dehydration of ethylene and ethyl on different surfaces.[22] Pd on gold is more reactive than pure Pd because the gold lattice constant is larger than that of Pd. Furthermore, Pd/Ru is less reactive because the Ru lattice constant is smaller. However, Pd and Re have essentially the same lattice parameter, but Pd/Re is nevertheless less reactive than pure Pd. In this case, the

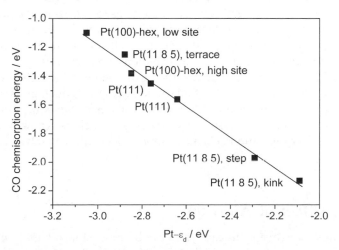

Figure 12 *Calculated chemisorption energy for CO atop Pt atoms in various environments as a function of the energy of the d-band centre*
(Reproduced with permission from Ref. 1)

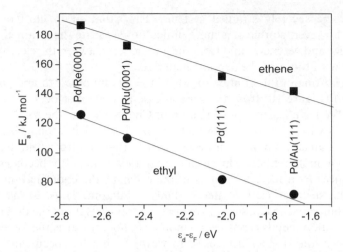

Figure 13   *Variations for the transition state energy for dehydrogenation of ethene and ethyl groups on Pd(111) and on palladium overlayers on various other metals.* (Reprinted from Ref. 22. Copyright (2000), with permission from Elsevier)

dominant effect cannot arise from strain but must be traced to the electronic interaction with the support.

Since adsorption on steps is energetically favourable over that on flat surfaces, a system may gain energy when it restructures by step formation as the coverage with a suitable adsorbate is increased. At high coverages, when the atoms below steps, which have a higher coordination and thus a lower binding energy to adsorbates than those on the flat surface, become populated then the trend is reversed and the steps heal out.[23] It was proposed that surface roughening as a function of adsorbate coverage is also a general phenomenon. Since reactivity increases with binding energy up to a certain extent, we must assume that not only the surface morphology changes as a function of gas composition and pressure but that there is a self-regulated change of surface activity by adsorption through the distribution of active sites.

It has also been observed that adsorbed hydrogen enhances the surface self-diffusion of Pt adatoms on a Pt(110) surface by reducing its activation barrier.[24] This has implications for many processes in which gas–surface interactions are important, such as sintering (which is known to depend crucially on gas composition), chemical vapour deposition, catalysis and crystal growth.

5   The Effect of Alloying

Traditionally, heterogeneous catalysts have been developed based on extensive trial and error experimentation, involving a large number of catalytic materials combined with some intuition. The group around the Danish theoretician Jens Nørskov has now demonstrated for the example of ammonia synthesis that tailored bimetallic catalytic materials can be obtained successfully by rational interpolation of the periodic table.[25] However, the principle is suggested to be applicable more generally.

The basis of the prediction is the well-known volcano-shaped relation of catalytic activity, often represented by the turnover frequency (TOF), as a function of adsorption energy (Figure 14). If the binding energy of the reactant to the catalyst surface is too weak, then it results in poor catalytic activity. If adsorption is too strong the reactant may transform into the product but this product does not desorb and blocks the catalyst surface so that its activity decreases. The optimum activity is obtained with an intermediate adsorption energy, which represents a compromise between a catalyst with a small activation barrier for the reaction that leads to a product-blocked surface and a surface with low nitrogen coverage.

For the reaction of N_2 with H_2 to form ammonia, the transition state is energetically (and in terms of bond lengths) very similar to the product. Therefore, the dependence of the catalytic activity on the nitrogen adsorption energy obeys the linear Evans–Polany relationship between the activation energy (which is N_2 dissociation) and the stability of surface-adsorbed N_2.[25]

Pd–Pt bimetallic-supported catalysts were found to be more sulfur resistant than each of the pure-supported metals.[26] The same is true for Fe–Pt catalysts.[27] Furthermore, it is well known that Ru–Pt alloy particles are much less susceptible to CO poisoning on fuel cell anodes. Theoretical work showed that the adsorbate-free surface of such an alloy particle has no Ru in the first layer, but the

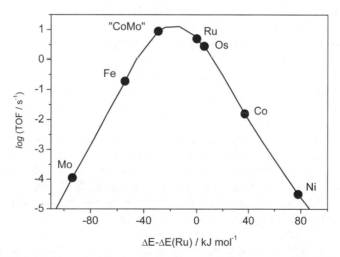

Figure 14 *Volcano-shaped plot of calculated turnover frequencies (TOF) for ammonia synthesis as a function of the adsorption energy of N_2. The point denoted 'CoMo' represents the predicted 1:1 alloy between Mo (which binds N too strongly) and Co (which binds N too weakly).*
(Reprinted with permission from Ref. 25. Copyright (2001) American Chemical Society)

presence of CO can move some Ru to the surface and all these Ru atoms are then covered by CO, while the Pt sites remain quite free from CO.[28] It was concluded that the superior performance of Pt–Ru compared to pure Pt as an anode catalyst can, at least in part, be attributed to a modification of the CO/Pt adsorption energy due to the presence of Ru in the bulk. Any good anode material must bind CO more weakly than pure Pt, but not too weakly, since this would imply that H_2 is also bound too weakly, which would render H_2 dissociation more highly activated, making it the rate-limiting step.

A more thermodynamic explanation starts from the fact that the energy of an atom in the bulk is different from that on the surface. In bimetallic alloys, the surface composition is thus in general quite different from that in the bulk. The element with lower surface energy will preferentially accumulate on the surface. Roughly, the surface segregation energy is proportional to the difference of the surface energies of the two alloy components. The surface energies, in turn, are proportional to the cohesive energies. For transition metals, the cohesive energies obey a parabolic behaviour as a function of the number of valence d-electrons with a maximum for half-filled bands and low values near the beginning and the end of a d-band series.[28] Thus, in bimetallic particles the central transition metals are usually confined to the bulk region while late transition metals such as Pt and Pd, which have promising catalytic properties, segregate to the surface.

The presence of adsorbates can change the surface composition of an alloy significantly. For example, CO adsorption energies grow approximately linearly with decreasing number of d-electrons, that is from late to early transition metals within a series. This means that in the presence of CO, there may be segregation reversal in the binary Pt or Pd alloys with earlier transition metals. The strong interaction with CO pulls the subsurface elements to the surface.

Recently, alloys have been discovered that exist only on surfaces and have no bulk counterpart. For example, Au is completely insoluble in bulk Ni. Nevertheless, Au atoms replace Ni in the first Ni(110) surface layer, forming a surface alloy. The squeezed-out Ni atoms agglomerate in Ni islands on the surface.[29] Experiments were supported by total energy calculations, which provided a detailed understanding of this amazing phenomenon. The minimum energy for Au surrounded by Ni is found at a coordination number of eight, because eight Ni neighbours at the Ni–Ni equilibrium distance contribute as much electron density around the Au atoms as twelve Au neighbours at the much larger Au–Au equilibrium distance. This means that in bulk Ni where the coordination of an Au atom is twelve the energy is high, whereas substituted into a Ni(110) surface where the coordination is seven the Au atom is close to its minimum energy.[29]

It was found that smaller particles were often less efficient in prominent reactions, giving rise to some disappointments when catalyst particle size effects were investigated, a prominent exception being gold. We remind that the adsorbate binding energy increases with decreasing particle size. For optimum catalysts, this leads to a shift away from the maximum on the volcano plot and thus to lower turnover frequencies.

6   Metal-Support Interactions

The art of catalyst preparation involves the preparation of a high surface area. The higher the area of accessible catalyst surface, the greater will be the number of product molecules per unit time. Many of the catalytic materials tend to sinter and are therefore difficult to prepare with a stable high surface area. Moreover, these materials are often precious metals, which have to be used in the most economical way. Therefore, many catalysts are prepared by distributing them onto some chemically inert support material that is available with a high surface area. Most commonly, these are inexpensive porous oxides, such as alumina, silica and zeolites, but also carbon, and more recently other oxides have been used.

Over time, it has become clear that the same catalytic material prepared the same way on different support materials does not always show the same performance. Apparently, a support does more than just supporting a catalytic particle; it may influence it rather strongly. There is a pronounced dependence of the electronic properties and the catalytic behaviour of noble metal particles on the acid–base property of the support. It is not trivial to separate support effects from those of all the other sources, and it is only relatively recently that this interaction has become a subject of systematic investigations.

We are well aware that there is a *thermoelectric potential* at the junction of two metals with different Fermi levels. In conducting materials, this interaction is associated with a partial electron transfer from the metal with a higher to that with a lower Fermi level. We also know that there is a dielectric response when we place an insulator between the two plates of a capacitor. Even though there is no electron transfer across the junction, in this case the dielectric can reduce the voltage on the capacitor by a considerable factor. We explain the effect by a polarisation of the dielectric material, which involves a limited displacement of electrons in atoms and in chemical bonds or the transfer of a number of electrons from the metal to the support or vice versa. The situation is akin to that of a *Schottky contact*. Furthermore, we are aware that at sufficiently low temperature, even two noble gas atoms can form a van der Waals bond and stick together. Again, we explain the effect by polarisation, which in this case is ascribed to dispersion forces.

A metal particle with a relatively low ionisation potential in contact with a support forms a junction with an insulator, which has a much higher ionisation potential. As long as the catalyst particle is metallic, it has a high dielectric constant, and it is plausible that there is a polarisation or redistribution of electronic charge in the particle. However, sufficiently small particles are no longer metallic, and it was recently demonstrated on the basis of vibrational spectroscopy of adsorbed CO on Pd and Pt, and on XPS and other electron spectroscopic techniques that simple polarisation is not sufficient to explain the effect.[30] Rather, a consistent picture requires that the energetic levels of the metal valence orbitals (the Fermi energy) change as a function of the metal–support interaction, resulting in an increase of the metal ionisation potential with increasing acidity of the support (Figure 15). It affects the chemisorption energies of adsorbates.[31,32] The origin of the effect was discerned as the Coulomb

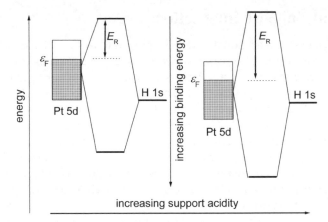

Figure 15 *Schematic molecular orbital diagram showing the downshift of the Fermi level of a Pt catalyst particle on an acidic support.*
(Reprinted from Ref. 30 with permission from Elsevier)

interaction between the metal particle and the support oxygen ions. Clear evidence of charging, based on shifts of C–O vibrational stretch frequencies and quantum chemical ab initio calculations, was also reported for bonding and catalysed oxidation of CO on Au_8 clusters on a F-centre-rich MgO surface.[33] In contrast, gold octamers on a F-centre-free MgO(001) surface were essentially inactive for the combustion reaction. The phenomenon reminds of the observations that the vibrational frequency of CO adsorbed on a metal electrode is also a function of the applied potential (see Section 6 of the present Chapter).

The above study illustrates the importance that a specific electronic interaction between catalyst and support can have. Other factors that may control reactivity and selectivity are the metal catalyst surface structure (steps, kinks and other defects) and diffusion of surface species between oxide support and metal (the so-called spillover). To conduct controlled experiments that test for the various contributions, electron beam lithography fabricated platinum nanoparticle model catalysts were prepared.[34] The reactivity of Pt/SiO_2 arrays consisting of metal particles with 28 nm diameter, 15 nm height and 100 nm square periodicity was compared with that of a Pt foil in cyclohexene hydrogenation and dehydrogenation reactions. In the dehydrogenation the turnover number per site was over three times higher on the nanoparticles than on the foil, and in the hydrogenation it was higher by more than 50% (Figure 16).

The electronic structure of the two materials was not considered to be different since each particle of the size in the array contains around 100,000 atoms. The primary difference between the Pt particle array and the foil was the presence of the $Pt–SiO_2$ interface. Thus, hydrogen spillover from Pt to the oxide surface could play a role as well as higher step and kink densities on the nanoparticle and in particular near the interface. Besides this, XPS studies on the effect of Ne sputtering on platinum thin film deposited on SiO_2/Si revealed platinum silicide formation at cracks on the Pt layer or Si support (Figure 17).

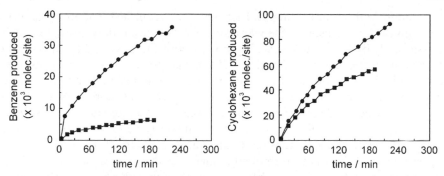

Figure 16 *Turnover numbers in molecules per site for benzene and cyclohexane produced from cyclohexene on the nanoparticle array (upper curves) and a Pt foil (lower curves).*
(Reprinted with permission from Ref. 34. Copyright (2001) American Chemical Society)

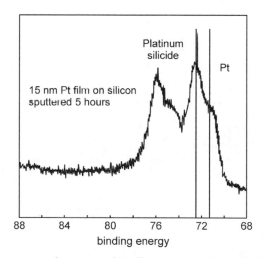

Figure 17 *XPS spectrum of a sputtered Pt film evaporated onto Si (100) wafer with 5–7 nm native oxide. The vertical lines mark the binding energies of platinum and of platinum silicide.*
(Reprinted with permission from Ref. 34. Copyright (2001) American Chemical Society)

It was suggested that silicide formation was also quite feasible under reaction condition, so that the true reason for the higher efficiency of the Pt nanostructure may be catalysis by Pt silicide.[34]

7   The Influence of an External Bias Voltage

It is a common practice in electrochemistry to change periodically the electrode potential to clean the electrode. This changes the availability of electrons to form bonds with adsorbate molecules at the catalyst surface. The effect can best be monitored by measuring the vibrational frequencies of the adsorbate, and a classic study by Zou and Weaver[35] reported both the C–O and the metal–CO stretching frequencies as a function of electrode potential. Figure 18 displays the results for CO adsorbed atop Pt atoms. Reminding that the vibrational frequency of a harmonic oscillator scales with the square root of the force constant, we conclude by inspection of the figure that a more negative potential strengthens the metal–CO bond and simultaneously weakens the C–O bond. Density functional calculations reveal an analogous picture for adsorbed O_2 on the surface of Pt.[36] A negative potential increases the electron density on O_2 in the direction of O_2^{2-} (in the absence of a potential, there is already partial electron transfer from the Pt support onto O_2) which goes along with a decrease

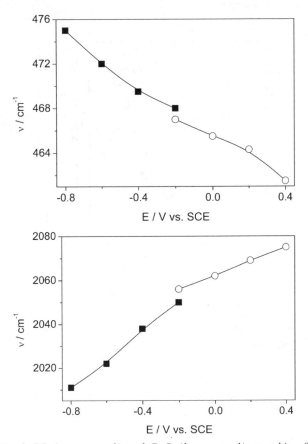

Figure 18   *Metal–CO (upper panel) and C–O (lower panel) stretching frequencies for CO atop Pt as a function of electrode potential. The filled and open symbols refer to alkaline and acidic electrolytes respectively.*
(Reprinted in part with permission from Ref. 35. Copyright (2001) American Chemical Society)

of the O–O stretching frequency and an increase of the O–O bond length, while the Pt–O distance gets shorter.

The vibrational absorption band for ^{13}CO adsorbed on Au_8 supported on F-centre-rich MgO(001) was shifted by 25–50 cm^{-1} to lower frequencies compared with F-centre-free MgO.[33]

Advantage has been taken of this effect for the improvement of the C–O tolerance of low temperature fuel cell catalysts by an electrode pulsing technique.[37] During the short periods where the anode potential was sufficiently positive CO was oxidised to CO_2 with the help of adsorbed water.

Also the ^{13}C NMR chemical shift of CO and CN chemisorbed on a 10 nm Pt electrocatalyst was shown to obey a linear relation with the electrode potential.[38] The more negative the potential, the larger the chemical shift. Thus, ^{13}C becomes more deshielded as the potential goes more negative, and the slope is about −71 ppm V^{-1} for ^{13}CO and −50 ppm V^{-1} for ^{13}CN on Pt. Even larger shifts were reported for adsorbates on Pd.

Oxide layers of 1–7 nm thickness grown on HF-cleaned Si(100) substrates were studied by means of XPS in the Si2p region.[39] The binding energy difference between the oxide Si^{4+} and the substrate Si^0 increases from 3.2 to 4.8 eV for samples containing a 1–7 nm oxide layer. External biasing with a d.c. source shifts the absolute position of XPS peaks to lower binding energies. More importantly, the chemical shift difference between oxidised and elemental silicon increases under negative bias, and it decreases under positive bias. This difference displays a sigmoid character around zero bias and reaches its saturation values of ca. ±0.2 eV within a few Volt bias voltages (Figure 19). The effect was attributed to differential charging between the silicon oxide layer and silicon substrate, which is decreased when a positive bias is applied to the sample.

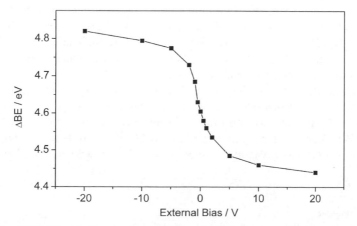

Figure 19 *XPS Si2p binding energy difference (ΔBE) between SiO₂ at a thickness of 4 nm and elemental Si support*
(Reprinted with permission from Ref. 39. Copyright (2001) American Chemical Society)

Key Points

- The surfaces of small particles are highly fluxional and adapt continuously and much more easily than single crystal surfaces to the changing adsorbates during catalytic reactions.
- The energetics of the HOMO and LUMO frontier orbitals can be tuned sensitively by the cluster size, which changes the availability of electrons to participate in bond formation to adsorbates or in redox reactions.
- Tensile strain leads to band narrowing, and more than half-filled d-bands of transition metals are pushed up in energy. In general, this leads to higher adsorption energies and to lower activation energies for catalytic reactions of adsorbates.
- Adsorption energies increase with decreasing coordination number of adsorption sites, that is in the order flat surface < steps < kinks < adatoms.
- Alloying and polarisation of the metal by its support as well as by an externally applied potential shifts the energy of the d-band and changes adsorption energies and reactivities of adsorbates.
- There are often a minimum number of atoms in a cluster that are necessary for two reactant molecules to adsorb in a geometry that permits the reaction to occur.

General Reading

- G.A. Somorjai, *Introduction to Surface Chemistry and Catalysis*, Wiley, New York, 1994.
- R.M. Lambert and G. Pacchioni (eds), *Chemisorption and Reactivity on Supported Clusters and Thin Films: Towards an Understanding of Microscopic Processes in Catalysis*, Kluwer Academic, Dordrecht, The Netherlands, 1997.
- B. Hammer and J.K. Nørskov, Theoretical surface science and catalysis – calculations and concepts; *Adv. Catal.*, 2000, **45**, 71.
- B. Hammer and J.K. Nørskov, Theory of adsorption and surface reactions, in *Chemisorption and Reactivity on Supported Clusters and Thin Films*, R.M. Lambert and G. Pacchioni (eds), Kluwer, Dordrecht, 1997.
- A. Wieckowski, E.R. Savinova and C.G. Vayenas (eds), *Catalysis and Electrocatalysis at Nanoparticle Surfaces*, Marcel Dekker, Inc., New York, 2003.

References

1. B. Hammer and J.K. Nørskov, *Adv. Catal.*, 2000, **45**, 71.
2. D.M. Cox, R. Brickman, K. Creegan and A. Kaldor, *Z. Physik D*, 1991, **19**, 353.

3. B. Hammer and J.K. Nørskov, *Nature*, 1995, **376**, 238.
4. S. Naito and M. Tanimoto, *J. Chem. Soc. Chem. Commun.*, 1988, **12**, 832.
5. U. Heiz, F. Vanolli, A. Sanchez and W.-D. Schneider, *J. Amer. Chem. Soc.*, 1998, **120**, 9668.
6. S. Vajda, S. Wolf, T. Leisner, U. Busolt and L. Wöste, *J. Chem. Phys.*, 1997, **107**, 3492.
7. I. Balteanu, O.P. Balaj, M.K. Beyer and V.E. Bondybey, *Phys. Chem. Chem. Phys.*, 2004, **6**, 2910.
8. M. Valden, X. Lai and D.W. Goodman, *Science*, 1998, **281**, 1647.
9. M. Mavrikakis, P. Stoltze and J.K. Nørskov, *Catal. Lett.*, 2000, **64**, 101.
10. H.G. Boyen, G. Kästle, F. Weigl, B. Koslowski, C. Dietrich, P. Zieman, J.P. Spatz, S. Riethmüller, C. Hartmann, M. Möller, G. Schmid, M.G. Garnier and P. Oelhafen, *Science*, 2002, **297**, 1533.
11. H. Häkkinen, M. Moseler, O. Kostko, N. Morgner, M.A. Hoffmann and B.v. Issendorff, *Phys. Rev. Lett.*, 2004, **93**, 093401.
12. A. Sanchez, S. Abbet, U. Heiz, W.-D. Schneider, H. Häkkinen, R.N. Barnett and U. Landman, *J. Phys. Chem. A*, 1999, **103**, 9573.
13. U. Heiz, A. Sanchez, S. Abbet and W.-D. Schneider, *J. Amer. Chem. Soc.*, 1999, **121**, 3214.
14. M. Daté, M. Okumura, S. Tsubota and M. Haruta, *Angew. Chem. Int. Ed.*, 2004, **43**, 2129.
15. F. Maillard, S. Schreier, M. Hanzlik, E.R. Savinova, S. Weinkauf and U. Stimming, *Phys. Chem. Chem. Phys.*, 2005, **7**, 385.
16. R. Narayanan and M.A. El-Sayed, *Nano Lett.*, 2004, **4**, 1343.
17. J.K. Nørskov, T. Bligaard, A. Logadottir, S. Bahn, L.B. Hansen, M. Bollinger, H. Bengaard, B. Hammer, Z. Sljivancanin, M. Mavrikakis, Y. Xu, S. Dahl and C.J.H. Jacobsen, *J. Catal.*, 2002, **209**, 275.
18. J.A. Rodriguez and D.W. Goodman, *Science*, 1992, **257**, 897.
19. E. Kampshoff, E. Hahn and K. Kern, *Phys. Rev. Letts.*, 1994, **73**, 704.
20. M. Gsell, P. Jakob and D. Menzel, *Science*, 1998, **280**, 717.
21. M. Mavrikakis, B. Hammer and J.K. Nørskov, *Phys. Rev. Letts.*, 1998, **81**, 2819.
22. V. Pallassana and M. Neurock, *J. Catal.*, 2000, **191**, 301.
23. P. Thostrup, E. Christoffersen, H.T. Lorensen, K.W. Jacobsen, F. Besenbacher and J.K. Nørskov, *Phys. Rev. Letts.*, 2001, **87**, 126102.
24. S. Horch, H.T. Lorensen, S. Helveg, E. Lægsgaard, I. Stensgaard, K.W. Jacobsen, J.K. Nørskov and F. Besenbacher, *Nature*, 1999, **398**, 134.
25. C.J.H. Jacobsen, S. Dahl, B.S. Clausen, S. Bahn, A. Logadottir and J.K. Nørskov, *J. Amer. Chem. Soc.*, 2001, **123**, 8404.
26. J.L. Rousset, F.J. Cadete Santos Aires, F. Bornette, M. Cattenot, M. Pellarin, L. Stievano and A.J. Renouprez, *Appl. Surf. Sci.*, 2000 **164**, 163.
27. J. Zheng, T. Schmauke, E. Roduner, J.L. Dong and Q.H. Xu, *J. Mol. Catalysis A: Chemical*, 2001, **171**, 181.
28. E. Christoffersen, P. Liu, A. Ruban, H.L. Skriver and J.K. Nørskov, *J. Catal.*, 2001, **199**, 123.

29. L. Pleth Nielsen, F. Besenbacher, I. Stensgaard, E. Laegsgaard, C. Engdahl, P. Stoltze, K.W. Jacobsen and J.K. Nørskov, *Phys. Rev. Lett.*, 1993, **71**, 754.
30. J.T. Miller, B.L. Mojet, D.E. Ramaker and D.C. Koningsberger, *Catalysis Today*, 2000, **62**, 101.
31. M.K. Onderhuijzen, J.A. van Bokhoven, D.E. Ramaker and D.C. Koningsberger, *J. Phys. Chem. B*, 2004, **108**, 20247.
32. D.E. Ramaker, M. Teliska, Y. Zhang, A. Yu. Stakheev and D.C. Koningsberger, *Phys. Chem. Chem. Phys.*, 2003, **5**, 4492.
33. B. Yoon, H. Häkkinen, U. Landman, A.S. Wörz, J.-M. Antonietti, S. Abbet, K. Judai and U. Heiz, *Science*, 2005, **307**, 403.
34. J. Zhu and G.A. Somorjai, *Nano Lett.*, 2001, **1**, 1.
35. S. Zou and M.J. Weaver, *J. Phys. Chem.*, 1996, **100**, 4237.
36. A. Panchenko, M. Koper, T. Shubina, S. Mitchel and E. Roduner, *J. Electrochem. Soc.*, 2004, **151**, A2016.
37. L.P.L. Carrette, K.A. Friedrich, M. Huber and U. Stimming, *Phys. Chem. Chem. Phys.*, 2001, **3**, 320.
38. Y.Y. Tong, A. Wieckowski and E. Oldfield, *J. Phys. Chem. B*, 2002, **106**, 2434.
39. B. Ulgut and S. Suzer, *J. Phys. Chem. B*, 2003, **107**, 2939.

CHAPTER 11

Applications: Facts and Fictions

1 Nanomaterials

1.1 General Considerations

Basically, nanomaterials have always existed and have been applied by humans. We remind of colloidal gold and silver particles which served as pigments in stained glass church windows and ceramics since the 10th century AD. Nanoparticles have been created for thousands of years as the products of food cooking, combustion, in volcanic activities and more recently from vehicle exhausts. Even nature has always taken advantage of the properties of nanostructures, for example of capillary forces or of the Lotus effect in plants. Nevertheless, the past decades have seen a large progress in synthesising and tailoring nanomaterials to order.

One of the main points to take care of is that nanomaterials are always thermodynamically unstable relative to materials consisting of macroscopic entities. This means that they have to be produced at sufficiently low temperatures where their growth is controlled kinetically and not thermodynamically. There has been great progress in the ability to control structures at smaller and smaller scales.

The strength and the durability of structural materials strongly depend on the structure and properties of grain boundaries. Therefore, grain boundaries will often have to be protected by thin passivating layers of oxides or organic molecules. Furthermore, interfaces usually act as barriers to dislocation motion and therefore strengthen materials with decreasing grain size, qualitatively described by the well-known Hall–Petch relation. This leads to special mechanical properties of nanocrystalline metals.[1]

Nanomaterials are to a large extent reality, but more detailed understanding of some basic principles is expected to extend greatly their potential for practical applications. Some of their promises will be outlined below.

1.2 Applications in Medicine

Nanoparticles offer some attractive possibilities in biomedicine. They are smaller or of comparable size to that of a cell (10–100 μm), a virus (20–450 nm) or a protein (5–50 nm). They can therefore move more or less freely in an organism. They may be coated to make them compatible with, or mimic, or bind to a biological entity of interest. Endothelian layers of fast-growing tumour tissue are often quite porous so that these particles can penetrate relatively easily. More specifically, some of the applications in actual or prospective use are as follows:

Liposomes are relatively stable micelle-like structures built from suitable amphiphilic block-copolymers or by self-assembly from phospholipids. They are already in use as submarine-like transport vessels or Trojan horses for *drug delivery*. Drug molecules may be intercalated in the lipophilic wall that mimics a cell membrane, or dissolved in the hydrophilic interior. The liposome protects the drug from being metabolised in certain environments, and due to its more hydrophobic character it may let it pass barriers between the digestive system and the blood or between blood and brain. When it arrives at the target site the drug may be released due to a different pH or an enhanced temperature in an inflamed region. The pH of tumour tissue typically assumes values of 4–5, significantly lower than the physiological pH 7.4. In lysosomes the pH can reach values as low as 4–5.[2] Acid labile linker molecules such as hydrazones are cleaved under these conditions so that the liposomes open up or protective shells are removed, permitting the drug to be released where it is needed. It has been demonstrated that significantly higher local concentrations can be reached at the site of interest, which increases greatly the selectivity and effectiveness of the drugs and furthermore reduces unwanted side effects.

Ferromagnetic nanoparticles have also been developed and are further optimised in view of applications for *targeted delivery of therapeutic drugs*, genes or radionuclides.[3] Generally, the magnetic component of these carrier particles (usually magnetite, Fe_3O_4, or maghemite, γ-Fe_2O_3) is coated with a shell of SiO_2 and then by a biocompatible polymer which can also be functionalised to provide attachment points for the coupling of cytotoxic drugs or bind antibodies to the carrier complex. A magnet is placed outside the body near the target site to capture the magnetic particles flowing in the circulatory system. This is expected to localise the action of cytostatic anti-cancer drugs and reduce unwanted side effects of anti-inflammatory drugs on patients who suffer on chronic arthritis. It should be noted that dextrane coated iron oxides are biocompatible and are extracted *via* the liver after the treatment.

Catabolism of tumours can be enhanced by *artificially induced hyperthermia*. One of the possible techniques involves dispersing magnetic particles throughout the target tissue and then applying an alternating magnetic field of sufficient strength (0–15 kA m^{-1}) and frequency (0.05–1.2 MHz) to heat them up. If the temperature of the surrounding tissue can be maintained above a therapeutic threshold of 42° C for 30 min or more then the cancer is destroyed.[3] The challenge is to deliver and localise an adequate quantity of magnetic particles, often assumed to consist of *ca.* 5–10 mg cm^{-3}, to generate enough heat.

In biomedicine, it is often advantageous to *separate out specific biological entities* from their native environment in order to prepare concentrated samples for analysis or further use. This may be achieved by tagging these entities with magnetic material and in a second step separate those out in a fluid-flow based magnetic separation device. Magnetic particles coated with immunospecific agents have been successfully bound to red blood cells, lung cancer cells, bacteria, urological cancer cells and Golgi vesicles.[3]

Another application of magnetic nanoparticles is that of *contrast enhancement agents* for magnetic resonance imaging. The effect of a magnetic contrast agent is to shorten the relaxation times, both T_1 and T_2. The uptake of nanoparticles by the tissue depends on their size. Larger particles with a diameter of 30 nm or more are rapidly collected by the liver and spleen, while those of 10 nm or less are collected in reticuloendothelial cells throughout the body, including those of lymph nodes and bone marrow.[3] Tumour cells do not have the effective reticuloendothelial system of healthy cells, so that their relaxation times are not altered by the contrast agent. This has been used to assist the identification of lymph nodes, liver tumours and brain tumours.

A quite different application of nanomaterials involves the coating of endoimplants with biocompatible layers which form a dense and mechanically stable interface between the biological tissue and the incompatible metallic materials that have otherwise desired properties and are easy to manufacture. These coatings often consist of ceramic type materials interfaced by self-assembled monomolecular layers which are fixed to the substrate by chemical bonds.

1.3 Intelligent Surfaces

Considerable efforts relate to the engineering of functional or even intelligent surfaces. Much progress has been achieved in making hydrophobic and dirt repelling fabrics and other surfaces, based on mimicking the Lotus effect, and strategies are now becoming available which allow these properties to be switched as a function of temperature or light exposure (see Section 4 of Chapter 6).

1.4 Applications in Catalysis

Nanomaterials are of utmost importance in catalysis. It has been explained in detail in Chapter 10 how reactivity and selectivity can be tuned by changes of particle size and shape, by introduction of deliberate disorder and surface fluxionality, and by alloying or support interactions.

1.5 Applications in Environmental Technologies

Nanomaterials have not yet been explored much in view of their potential use to solve environmental problems, but a number of applications are under way.

Nanoporous aluminosilicates (zeolites) are already being used as adsorbents to bind radionuclides and poisonous transition metal ions from waste water. Their advantages consist in their high sorption capacity and the tuneable selectivity.

Nanoporous membranes are based mostly on zeolites or on tailored polymer membranes containing sintered nanoparticles with adjustable pore size. They are used to filter microdust particles out the air and of water, and they do this more efficiently than traditional filter materials. Furthermore, they permit the inclusion and immobilisation of catalytically active metal oxide centres or of biocatalysts to directly degrade problematic pollutants such as chlorinated ethenes. Possible applications include the preparation of potable water and the separation of viruses and bacteria.

Catalysts for automotive exhausts or fuel cell applications have been known for a long time. Prominent new materials include gold nanoparticles which are used for the degradation of toilet odours. Some of the materials are photocatalysts and need activation by light. Titanium dioxide is used as coating for the shielding of electrical lights in tunnels. It converts deposited soot particles to CO_2 and keeps the shielding transparent. In a similar way zinc oxide serves as a photocatalyst for the degradation of chlorinated phenols. In near future, the walls of swimming pools will be coated by an ultra-thin layer of photocatalytically active materials. Activation of oxygen leads to the formation of hydroxyl radicals which are strong anti-bacterial agents and can degrade many organic compounds.

Sensor technology has made tremendous progress in recent years. A multitude of nanobased sensor materials (metals, semiconductors, polymers) is deposited on selected areas of a chip or on nanocantilevers and serves to analyse gas mixtures by measuring solubility, vapour pressure, melting point, boiling point or heats of adsorption or reaction of the components. Other sensors are functionalised by biomolecules which allow a selective trapping of biomolecules, viruses or cells *via* molecular recognition. Such a complex miniaturised sensor which is often termed *lab-on-a-chip* requires a high degree of interdisciplinary cooperation between chemists, biochemists, physicists, materials scientists and engineers and is an excellent example of what can realistically be expected from nanotechnology.

2 Nanotechnology

While applications of nanomaterials take advantage of the special properties of normally macroscopic amounts of nanoscopic materials, nanotechnology aims at engineering, addressing and manipulating individual nanospecimens. This is a much larger challenge, and this is mostly what Feynman suggested in his legendary lecture. In Drexler's view, nanotechnology is *molecular manufacture*. Over the last decade, scientists have made significant progress in this direction. However, the experience is that the periphery needed to administrate even a single nanodevice tends to be a formidable megamachine. This is a fact, but if it is understood as a negative criticism then it misses the point. If we look back at three decades of developments of conventional computers then we have to acknowledge that the volume of the periphery has been reduced by a large factor while at the same time the computing power has increased tremendously. More importantly, the expectations are at least in part that nanodevices are not only smaller but that they can do fundamentally new things.

2.1 Applications of Nanomechanics

In his 1986 book *Engines of Creation*, Eric Drexler speculated that self-replicating robots with molecular-scale dimensions could be used to build everyday goods cheaply and efficiently or which could be distributed through the blood stream to undertake nanoscale surgery at places which are difficult to access.[4] He also warned that such *nanobots* could be dangerous if they escaped. The idea got a lot of influence, but it has meanwhile become clear that this scenario is unabashed hype. Eric Drexler himself acknowledges that nanoscale manufacturing does not require self-replicating devices.[5]

Nevertheless, molecules are being developed, which can change conformation upon stimulation by temperature, pH, light or electron transfer and act as nanomanipulators, nanomuscles or for the storage of information. Some of the concepts are based on rotatory or linear motion of threaded catenanes or rotaxane dimers.[6,7]

Recently, a nanoscale electromechanical rotor consisting of a 300 nm rectangular gold blade that sits on a multiwall nanotube axle was reported.[8] The ends of the axle were embedded in electrically conducting anchors that rested on the oxidised surface of a silicon chip. The rotor plate assembly was surrounded by three fixed stator electrodes. It was demonstrated by scanning electron microscopy that the metal blade changed its orientation as a function of the electrical voltage applied to the stator electrodes. Possible applications envisaged for this device include deflectable mirrors or sensor paddles for the detection of the motion of microfluidic systems.

Another report provides evidence of what the authors call a thermally driven single-molecule nanocar with a fullerene-based wheel-like rolling motion, not stick-slip or sliding translation.[9] Since the 'cars' move in a direction perpendicular to their axles the authors conclude that they move by a rolling motion of the fullerene wheels which rotate about the carbon–carbon bonds of the axles.

Many applications of nanomaterials are based on carbon nanotubes (CNTs). They are mechanically very strong, with a Young's modulus of over a terapascal which makes them as stiff as diamond with respect to stretching while flexibility about their axis is retained. Depending on the helicity of the graphene sheets they are semiconducting or metallic.[1] In a hot-drawing process inspired from textile technology wet-spun composite fibres consisting of single- and multiwall CNTs and poly(vinylalcohol) with a diameter of 30 μm and a length of 10 m can be produced.[10] Several hot-stretched fibres were tested, and values of tensile strength between 1.4 and 1.8 GPa were found. They hold great potential for a number of applications such as bulletproof vests, protective textiles, helmets and so forth.

Impressive nanofabrication capabilities have been demonstrated with scanning probe manipulation of atoms and molecules. While these techniques can manipulate atoms already present they have a loading deficiency and cannot efficiently deliver atoms to the work area. In this context, a highly creative experiment led to the suggestion to use CNTs as nanoscale mass conveyors.[11]

The surfaces of CNTs were decorated with metallic indium and placed on the stage of a transmission electron microscope to monitor the process by video imaging. Individual CNTs were then contacted with conducting tungsten tips. Application of a voltage between these tips and the sample holder established an electrical current through the tubes and introduced thermal energy into the system *via* Joule heating. By increasing the voltage the local temperature can be increased past the melting point of the indium particles decorating the tubes. At a given temperature, an indium drop maintains an equilibrium indium surface concentration on the host CNT at its attachment point. Depending on the sign and magnitude of the dc bias voltage, drops near one end of the tube grow or shrink quite dramatically at cost of drops near the other end of the tube. Reversing the sign of the voltage reverses the process. Application of a few volts between the two tube ends can transport the metal efficiently over distances of several nanometres. The transport process has similarities to conventional electromigration, but several other mechanisms are also considered and cannot be excluded at this point.

Nature itself has devised various nanomechanical elements. Spinning flagellae bundles reminiscent of small propellers permit bacteria to move in one direction or the other. The kinesin motor protein is able to move along microtubules with control of directionality, carrying and transporting various components of the cell over distances of micrometres.[12] The ATP synthase is *nature's smallest rotary motor* and provides most of the chemical energy that aerobic and photosynthetic organisms need to stay alive.[13] The sense of rotation of the central subunit of the ATP synthase protein was demonstrated by attaching magnetic beads and using an electromagnet to rotate the beads. Clockwise rotation caused an increase in ATP, while ATP consumption was related to counter-clockwise rotation.

Over the past decades the *machine* metaphor has invaded the language of biologists. Each entity of a living cell is described as a machine: ribosomes are assembly lines, polymerases are copy machines, proteases and proteosomes are bulldozers, membranes are electric fences and so on. Although biologists generally agree that living systems are the product of evolution rather than of design, they describe them as devices designed for specific tasks.[14] A major objective of nanotechnology is to build nanomachines that are more efficient than conventional machines. They are self-assembling, self-curing, self-replicating and they adapt their production to the actual needs. The DNA–RNA system provides the code and the instructions for the machine to operate. Indeed, mimicking cellular nanomachines is a marvellous challenge. The magical recipe of building them by self-assembly is hoped to be extremely simple: take the proper molecular building blocks, pore them together in a beaker, stir and they will fall together into ordered, functioning entities without the need of being manipulated by any tiny construction robots. For comparisons, just imagine that you build a watch or an automobile by taking the parts, put them in a box and shake, and they will assemble to the working product. Over the past several decades the manufacture of cars and watches has indeed been changed from assembly by a multitude of workers to a nearly fully automated process that runs with very little human interference, but using sophisticated robots. What is needed now is to shrink the huge production roads and halls – and the products – by six orders of magnitude or more, and to find a strategy that does no longer need any robots!

Spontaneous surface alloying is driven by free energy, which at surfaces means surface tensions. When small amounts of tin are deposited on a Cu(111) single crystal surface at 290 K one observes that the tin crystals proceed to move spontaneously along the surface in a systematic fashion, leaving bronze alloys in their tracks.[15] Sn atoms within the islands randomly exchange with Cu atoms in the surface. The exchanged Cu atoms are ejected from the Sn islands in the form of an ordered 2-D bronze crystal. The motion is driven by the lowering of the surface free energy by alloying. When the islands meet a track they do not cross, instead they change direction if this allows them to move over an unalloyed region of the surface. The driving force of the motion could be used to drive nanomotors.

2.2 Applications in Nanoelectronics

The best way to increase the speed at which electric charges work is to decrease the distance they have to travel. This is the principle that electronic engineers have followed over the past four decades. In 1965 Gordon Moore, one of the founders of the Intel microchip company, predicted that the number of transistors per unit area on a microchip will double every year.[16] The prediction was found to hold roughly until today, although at the somewhat slower pace of doubling every 18–24 months. Chip manufacturers believe that this pace can be kept up for the next decade, but there is an unavoidable limit when the space necessary to store one bit hits about 4 nm, when troublesome quantum behaviour starts to pop up. Furthermore, heat will become a huge problem, and neighbouring bits would interact so that the independence of bits would be lost. This would happen in 13 years, assuming strict adherence to Moore's law Figure 1.

In his famous 1959 talk Richard Feynman estimated that 'all of the information that man has accumulated in all the books in the world, can be written in a cube of material one two-hundredth of an inch wide'.[17] This corresponds to a cube of $5 \times 5 \times 5 = 125$ atoms to store one bit, which is comparable to the 32 atoms that store one bit in DNA. In the meantime, miniaturising electronic devices has made great progress, and near atomic scale memory at a silicon surface has been reported.[18] Thereby, a bit is encoded by the presence or absence of a silicon atom inside a unit of $5 \times 4 = 20$ atoms, with the remaining 19 atoms required to prevent adjacent bits from interacting with each other. Readout of such a memory is *via* scanning tunnelling microscopy (STM), and writing is by removing individual atoms from the surface to write a zero. The memory is preformatted with a 1 everywhere by controlled deposition of silicon onto vacant sites. This approach increases the storage density from that of magnetic hard disks (16 Gbit cm^{-2}) to more than 10^4 Gbit cm^{-2} at cost of several orders of magnitude in the readout rate.[18] The stability of an individual

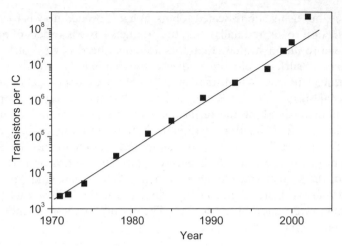

Figure 1   *Number of transistors on an integrated circuit between the Intel 4004 processor
(1971) and the Itanium 2 processor (2002). The line is for a doubling of the
number in 24 months (Moore's law)*

bit at room temperature was estimated to amount to 2–3 years, given by the
thermal stability of a Si adatom.

What exactly the building blocks of future nanoelectronics will be is an open
question. Numerous solutions for devices which are hoped to be able to replace
conventional ones are being investigated. Reliability and inexpensive production techniques will be essential features for large scale use. It will be difficult to
make them by self-assembly alone, but in combination with nanolithographic
and perhaps nanomanipulation techniques self-assembly comes in when thin
layers have to be deposited or when nanowires are grown by decoration of the
edges of a stepped surface.

The most advanced nanomaterials for applications in electronic devices are
organic thin films. Already in commercial production are high-efficiency, very
bright and colourful thin displays based on organic light-emitting devices
(OLEDs).[19] Significant progress is also being made in the realisation of thin-
film transistors (TFTs, see Figure 4 of Chapter 4) and thin-film organic
photovoltaic cells. The principal interest stems from the ability to deposit
organic thin films on very-low-cost substrates such as glass, plastic or metal
foils using techniques such as spray or spin coating, vapour deposition and ink-
jet printing. The aim is not primarily to find replacements for conventional
high-performance silicon electronics. Rather, due to their unique properties, for
example their softness, organic materials can fill niches not occupied by Si-
based devices. While the carrier mobility in inorganic materials typically ranges
from 100 to 10^4 cm^2 V^{-1} s^{-1} at room temperature it is at best on the order of 1
cm^2 V^{-1} s^{-1} for organic materials, depending on the packing order.

The ability to measure the conductance of a single molecule wired to two
electrodes is a basic requirement toward electronic devices based on single
molecules. Measurement of single-molecule conductance as a function of pH

was recently demonstrated for three peptides of different lengths.[20] Conductance was shown to decrease exponentially with length, suggesting a tunnelling mechanism and the corresponding tunnelling barrier height decreased with pH.

Single wall CNTs have an extremely high potential in view of applications in nanoelectronics. Their electrical conductivity depends on their diameter and the helicity of their graphene sheet which is described by the chirality vector (n,m). The 'armchair' versions which are rolled so that the C–C bonds are perpendicular to the tube axis ($n = m$) are metallic, whereas the zigzag versions with the C–C bonds parallel to the tube axis (n,0) are semiconducting. Intermediate forms are chiral and are metallic if (n–m) = $3i$, with $i = 0, 1, 2, 3, \ldots$, and ultrasmall CNTs with a diameter of 0.4 nm even exhibit superconductivity below 20 K.[21] When the CNT grows with a kink the chirality vector and thus the electrical properties change at the kink. This leads to a junction between a metal and a semiconductor or between two different semiconductors, which can be used as a diode. Logic circuits with field-effect transistors based on CNTs have been demonstrated.[22] These properties, taken together with their extremely high tensile strength and mechanical flexibility, make CNTs key elements for various future nanoelectronic devices. For example, an extreme sensitivity of their electrical resistance to gaseous molecules such as O_2, NO_2 or NH_3 makes them promising elements as chemical sensors.[23,24]

CNTs can be grown on a surface as arrays of upright rods. Because they are electrically conducting electrons can move freely up and down the tubes. This means that they can serve as aerials for electromagnetic radiation of the wavelength on the order of the length of the aerial, in this case for visible light.[25] It was verified that the tubes can function as aerials when the oscillating electrical field was oriented parallel to the tubes but not in a perpendicular geometry. This means that an array of CNTs could convert the light beams inside an optical computer into electrical signals, providing an interface with conventional electronics.[25]

Arrays of vertically aligned individual CNTs can work as cold cathode field emission electron source that can operate at high frequencies and high current densities. Because of their low weight, their instantaneous response and since they do not require heating they should prove valuable for miniaturised microwave devices in telecommunication in satellites or spacecrafts.[26]

Electronic elements have to be wired up. Unlike ordinary wires, *nanowires* are quantum devices. The electronic conductance in a nanowire can assume only discrete values in multiples of the *Landauer constant*, $2e^2/h$ (e is the electron charge and h is Plancks' constant). Furthermore, nanowire conductivity is strongly influenced by edge effects and therefore increasingly sensitive with decreasing diameter to defects at the surface. CNTs may be used as nanowires, or chains of metal atoms (e.g. by decorating the edge of a stepped silica surface with metal atoms) or of semiconducting materials.

Further indispensable elements of every electronics are switches. A robust and highly reliable *nanoswitch* between two wires, one from platinum and the other one from silver sulphide (Ag_2S) which is a solid electrolyte, was reported recently by Terabe *et al.*[27] In the open state, the two wires are spaced about

1 nm apart. This is controlled by applying a positive bias voltage to Ag_2S, which causes a silver nano-protrusion to form, creating a bridge between the two wires. Application of a negative voltage causes the silver bump to shrink, so that the atomic bridge is broken and the switch reopened. It worked more than 10^5 times both in air and in vacuum, at a rate of up to 1 MHz. Furthermore, by using pulsed bias voltages to tailor the length of the protrusion it was possible to switch between four different levels of quantum conductance.

Quantum computing is one of the terms that are loaded with high expectations. It involves addressing and manipulating individual quantum bits (*qubits*). It is the qubit rather than the principle of quantum computing that relates to nanotechnology. A qubit is a physical entity that can exist in two or more quantum states, as for example an artificial atom (quantum dot), a defect in a solid that contains an unpaired electron, or a molecule which is built up from well distinguishable magnetic nuclei. An important property of a qubit is its isolation from its physical environment so that its quantum states achieve the long decoherence times needed for successful computation. Examples are endohedral atoms in fullerenes, $N@C_{60}$ and $P@C_{60}$. Due to the absence of significant chemical interactions of these atoms with the fullerene cage and the high symmetry of this cage the decoherence times are very long.

2.3   Applications of Single Spin- and Nanomagnetism

Because of their small size, non-interacting, highly anisotropic molecular magnets are of potential interest for information storage, promising a very high information density. A significant problem that needs to be overcome in this context is the low temperature (< 4 K) that is currently needed to stabilise a given spin state. Furthermore, addressability and manipulation of individual molecular magnets need to be solved.

The operation of traditional semiconductor electronic devices depends on the transport and storage of electric charge. However, electrons have spin as well as charge, and adding the spin degree of freedom may revolutionise electronics, leading to new devices such as spin transistors, spin memory storage and even spin quantum computers. The advantages of these new devices would be non-volatility, increased data processing speed, lower electrical power consumption and increased integration densities compared with conventional semiconductor devices. The new technology is called *spintronics* (spin transport electronics or spin based electronics).[28]

Charge and spin are both properties of the electron. This does not mean that they transport at the same velocity. It is because the charge density is the sum of electron densities with spin α and spin β, $\rho(\alpha) + \rho(\beta)$, but the spin density is the difference of the same two properties, $\rho(\alpha) - \rho(\beta)$. We can imagine that an electron of spin α exchanges with an electron of spin β in the same sample. This changes the spin distribution but not the charge distribution. It thus represents a spin current but not a charge current. On the other hand we can imagine a pair of α and β electrons to move simultaneously, which leads to a charge current but not a spin current. In general we may have a situation between these two extremes, and recent investigations revealed that spin lags behind charge in a GaAs semiconductor sample.[29,30]

One of the motivations for the interest in spin electronics originates from the discovery that alternating thin layers of non-magnetic materials and ferromagnetic metals show large changes in their electrical resistance when a small magnetic field (100–1000 Oe) is applied. This change is known as *giant magnetoresistance* (GMR) and can amount to more than 60% at room temperature. The resistance of the material is lowest when the magnetic moments in the ferromagnetic layers are aligned parallel and highest when they are antiparallel. A spin current is created from the movement of spin-up electrons moving in one direction and spin-down electrons in the other. GMR spin valve read heads are already dominating applications in hard disc drives.

The magnetic fields necessary for the GMR effect are difficult to create in small electronic devices. It was therefore a breakthrough when it was found that electrons driven by a voltage through a thin semiconductor channel produces a perpendicular flow of spin current across the channel.[31,32] As a consequence, spin accumulates at the edges of the semiconductor channel in what is called *spin Hall effect* (SHE). Unlike the conventional Hall effect and the quantum Hall effect the SHE does not produce a transverse voltage nor does it need an external magnetic field. The polarisation relies instead on the relativistic coupling of the electrons spin and orbital angular momentum. The new technique opens an important new pathway towards solid state spintronics.

In today's electronics, logic operations take place in semiconductors while information is stored in magnetic media. It would be attractive if it were possible to combine both functions on the same chip, or even in the same bit. One of the remaining challenges of spintronics using devices which consist of a combination of ferromagnetic and semiconducting materials is the injection of spin-polarised carriers (electrons or holes) into a semiconductor, but various ways to overcome the problems are being investigated successfully.[28]

2.4 Application of Optical Properties

Soap bubbles or thin oil films on water appear differently coloured depending on the incident angle due to interference of reflected white light on both surfaces of the film. The beautiful colours of the wings of butterflies, the eyes of insects or the feathers of peacocks are based on the same principle: light is reflected and interferes from nanostructures with a periodicity on the order of the wavelength of the reflected light.

The principle of generating colour through structure in place of dye molecules is the basis for *photonic crystals*. With this aim, periodic structures with an accurate periodicity of the order of 400–800 nm have to be synthesised. This can be realised in three dimensions in arrays of monodisperse nanoparticles which are imbedded in a medium that has a diffraction index sufficiently different (by *ca.* 0.2 units) from that of the particles in order to generate diffraction effects. Such structures can be synthesised with good precision on the basis of core-shell nanoparticles. The colour depends on the angle of observation. Commercial applications of such materials are in the area of wrapping foils and decoration paper, and even in hair sprays and nail polish. One-dimensional crystalline arrangements arise from the lamellar structures of cholesteric liquid crystals.

A very intriguing application of photonic crystals relates to the design of materials with *negative refractive index*. This has the effect that light is 'bent the wrong way',[33] positive lenses would become negative, flat sheets could focus, the *Doppler effect* would be reversed, and other counterintuitive phenomena would occur. The refractive index of a medium is $n = \pm \sqrt{(\varepsilon\mu)}$. Conservation of energy requires that the positive sign of the square root is used for conventional positive dielectric permittivity ε and magnetic permeability μ, thus $n > 0$. However, if both ε and μ were negative the correct solution is that with the negative sign, and thus $n < 0$. Materials with only one of the two properties < 0 do not transmit light.

Meanwhile, a number of photonic crystal materials have been designed and tested. Shelby et al.[34] designed a two-dimensional periodically microstructured circuit board with split ring resonators to give negative μ and fine copper wires for negative ε. That the negative index of refraction allows a flat slab of material to behave as a lens and focus electromagnetic waves to generate a real image was demonstrated for an array of cylindrical alumina rods.[35] The value of the effective index of refraction depends on the angle of the incident wave, and a sharp image is only seen for a narrow frequency band. A particular advantage of the photonic crystalline material is its scalability to submicrometre dimensions for possible applications from microwave to optical frequencies.

Semiconductor lasers are key components in many widely used technological products, including compact disk players and laser printers, and they play a critical role in optical communication schemes. Because zero-dimensional semiconductors, or quantum dots, have discrete energy states rather than a continuous distribution (see Figure 2 of Chapter 4) they have theoretically predicted advantages that include temperature insensitivity, lower lasing thresholds, and a gain profile concentrated into a much narrower spectral range than the corresponding bulk material.[36] This has been confirmed in recent works using a thin-film composite of CdSe nanocrystals in a titania matrix prepared by wet chemistry.[36]

Certain cosmetic products contain ZnO_2 nanoparticles which act as UV absorbers and serve as protective solar lotions. In other products, TiO_2 particles of around 500 nm diameter are added to polyamide fibres which form the basis of UV absorbing fabrics with a protection factor of up to 80.

Attractive technological applications of light absorbing nanomaterials are thin-film solar cells. The energy is absorbed in the form of visible light and converted to electrical energy *via* a crucial charge separation process. The absorbing layer can be based on inorganic semiconductors with a band gap that corresponds to the energy of the absorbed radiation and makes optimum use of the solar spectrum. Alternatively, the light can be absorbed by organic dye molecules with high absorption coefficients. They mimic the chlorophyll molecules in leaves where charge separation is also the key step. When the absorption band is narrow so that only a fraction of the spectrum is absorbed one can take a multilayer approach in which several cells with different absorption characteristics are stacked. Besides this, some of the challenges in building these electrical devices include the synthesis of largely defect-free layers without any short-circuits, and the minimisation of charge recombination which would reduce the cell efficiency dramatically.

Thomas Edison fabricated electrical light bulbs based on black-body emission of electrically heated carbon filaments derived from bamboo fibres which were placed in a high-vacuum enclosure made of glass. Unfortunately, the carbon filaments were very fragile and the bulbs darkened rapidly as a result of vapour deposited carbon on the glass surface. Decades after carbon filaments had been replaced by tungsten it was suggested recently to return to carbon and use single- or double-walled nanotube filaments.[37] Tests revealed that nanotubes exhibit a lower threshold voltage and higher brightness at the same voltage when compared to the conventional safelight. The emitted light spectrum at high temperature is a combination of black-body radiation and electroluminescence so that at the same temperature a higher fraction of the electrical energy is emitted in the visible spectrum. Investigations of the durability revealed that the nanotubes showed great stability and worked well even after being turned on and off for more than 5000 times. In addition, if cleaned from amorphous carbon and residual catalyst, they could withstand continuous operation at 25 V ($\sim$1400 K) for more than 360 h without any visible evaporation.

3 Hopes, Hazards and Hype

3.1 Is Nanotechnology Useful?

Governments and companies would like to make sure that final benefits will be higher than investments. They want to make profit or make responsible use of tax payers' money. Nobody wants to blame them for this. But is money all that it is about? Is wealth creation the only valid goal of our life? How useful is it to watch a football game? What is an artistic sculpture in our garden good for? What is the benefit of going to a concert? We do not normally ask questions this way because we all know that there are things in our lives which are beyond our absolute needs for physical survival – but we would not want to miss them. We call them human culture. Science, just as well as arts, literature, or sports, is part of human culture and identity. We do not do it only for money. We want to learn what it is that drives the world. Usefulness and wealth creation are not a sufficient measure of the quality of life.

Nevertheless, there can be no doubt that products of nanotechnology will play an increasing role in our every day's life in future. We will list and evaluate briefly a number of instructive examples in the following sections, trying to separate facts from fictions, to illustrate what researchers are dreaming up and what they have already accomplished or what shows realistic promises.

3.2 Potential Health or Environmental Hazards

Nanotechnology will produce many novel products for application in every day's life. There is no general hazard that is obvious at this point, but they need to be evaluated carefully in the same way as any conventional product. We do not yet know how cells and tissue react to nanoparticles. Many of them will be used in an embedded form so that there is no direct exposure. However, when used in cosmetics or in medical applications, or when dust particles from mechanical wear of materials or from air pollution are inhaled we are reminded that conventional materials of similar constitution may be highly problematic (*e.g.* micrometre size quartz particles, asbestos fibres or smoke and soot). Concerns have been expressed that the very properties that are being exploited by researchers and industry, such as the high surface reactivity and ability to cross cell membranes, might have short or long term negative health and environmental impacts, and particularly, that they might result in higher toxicity.[1] For example, it remains to be seen whether there could be any toxicological problem due to the photoactivity of titanium dioxide nanoparticles which are being used in some sunscreens because they are transparent to visible light while acting as absorbers and reflectors in the ultraviolet.

The exposure to nanoparticles can also occur *via* indirect routes through pollution of the environment followed by uptake through food or drinking water, or it can represent a risk to the environment itself. Possible current sources, among others, may be the waste streams from factories or research laboratories. As a precaution, the environmental persistence, toxicity and bioaccumulation of the nanoparticles will have to be assessed.

It is important to put the concerns in context by noting that humans have always been exposed to some types of nanoparticles arising from natural sources such as atmospheric photochemistry or forest fires, resulting in millions of pollutant nanoparticles per breath, and that other types of nanoparticles have been in extensive use already for some time without significant problems, for example in tooth paste, sun lotion, paint or toners of printers and copy machines. This does not mean that the possible hazards arising from exposure to new types of nanoparticles should not be assessed and controlled carefully.

3.3 Ethical and Social Threats from Nanotechnology

Most of the ethical and social issues arising from applications of nanotechnologies will not be new or unique to nanotechnologies. Rather, they reflect the rapidly changing society due to the impact of the technological developments which we have experienced over the last century in general. It is well known that such developments create both winners and losers, for example as employment is displaced from one sector to another. It is difficult to predict whether there will be any breakthrough which will accelerate certain developments to an extent that leads to a new class of problems. Even though nanotechnologies may provide cheap sustainable energy, environmental remediation, radical advances in medical diagnosis and treatment and in improved consumer products which in principle could be of great benefit to poor people and developing countries, the experience hitherto has been that in reality such developments tend to increase rather than decrease existing divisions between the rich and the poor.

There is no way to avoid that nanotechnology will also be used for military purposes. Again, most of these issues are not specific to nanotechnologies. The line between non-military and military industrial activities may be hard to draw, and technologies are potentially available to a very wide range of parties, including individuals or small groups. An example that has received much publicity in the US is the threat by anthrax, a mass destruction weapon which was brought outside a biological weapons laboratory and distributed by public mail. Its efficiency relies significantly on special technologies for subdivision into ultra-fine particles. The manipulation of further biological and chemical agents using nanotechnologies could result in entirely new threats that might be hard to detect and counter. The high sensing and data processing capabilities will improve the efficiency and flexible use of weapons. On the other hand, the large number of cheap and selective nanosensors will improve defence capabilities through early detection of chemical or biological releases and increases the surveillance capabilities. Nanotechnology will contribute to create light weight and comfort battle suits which may make soldiers less visible and less vulnerable to enemy and environmental threats and at the same time focus on injury intervention and cure.[1]

A significant socio-ethical threat may derive from the efficiency of computing, telecommunication and sensing technology which permits to collect, store, combine and evaluate huge amounts of data. The progress in this sector will be tremendous and influence many aspects of our lives. The easy way in which human individuals can be identified, characterised and their mobility traced may be a chance for fighting organised crime and terror. The back side of the medal is that such surveillance threatens our traditional desire of privacy, and there is some concern that those who monitor use the information also for purposes which are different from those declared, and that the tools may be used to install a new quality of unilateral political and economic power.

The public awareness is currently mostly dominated by the exciting nature and untapped potential of nanotechnologies. More than in the issue of gene technology it is felt that nanotechnology is a natural progression of technology which is partly based on principles which have been applied by nature for a long time, and learning from nature cannot be so wrong (Lotus effect). Nevertheless, some of the involved parties are anxious to avoid a public debate like the one about genetically modified crops.[38,39] In view of this situation it is important to draw a realistic rather than an exaggerated picture of the prospects and to promote a transparent public dialogue with experts. Furthermore, it is recommended that governmental regulatory bodies review the adequacy of their procedures to ensure that any regulatory gaps are identified at an appropriate stage when new short or long term hazards appear at the horizon.

3.4   Is Nanotechnology a Hype?

Nanotechnology has often been charged to be nothing but hype. Let us be honest. The field is certainly full of exaggerated claims. But nobody can, at this point, judge seriously the outcome of all the claims. Only history will tell us which of the claims have turned into facts.

However, conducting the debate this way misses the point. The relevant fact is that fascination by nanotechnology has set free huge amounts of energy – and money – and initialised a wealth of creativity and imagination. Thousands of scientists dream the impossible and work hard to turn these dreams into reality. It is this collective brain power, which keeps things moving, and the outcome may be in many ways different from what experts think today. We may therefore conclude: don't mistake the claims for facts, but give creativity a chance!

Key Points

- Nanomaterials are always thermodynamically unstable compared with bulk; they therefore have to be synthesised under conditions of kinetic control at sufficiently low temperature.
- Medical applications of nanomaterials include drug delivery in liposomes, targeted drug delivery using ferromagnetic carriers, contrast enhancement and hyperthermic tumour treatment.
- Nanomaterials find applications in environmental technologies as adsorbents to bind radionuclides, as filters for microdust particles, for the immobilisation of catalysts, in photocatalysis and in sensor technology.
- Applications of nanomaterials are much more advanced than those of nanotechnology.
- Carbon nanotubes hold great promise as mechanical fortifiers and as versatile elements of nanoelectronics.
- Spintronics, the combined use of spin and charge of the electron, is hoped to lead to new electronic devices of small dimension, high speed and low power consumption.
- Photonic crystals are periodic nanostructures imbedded in a medium of sufficiently different index of refraction. Suitable conditions may lead to materials with an effective negative index of refraction.
- Small size quartz particles, asbestos fibres, smoke and soot are highly problematic for health. There is therefore an urgent need to pay attention to possible health hazards of nanomaterials which may be inhaled or applied on the skin.
- Nanotechnology has the potential to change the world. There will be new profiteers and new losers which may lead to significant socio-ethical threats.

- Is nanotechnology but hype? Let's take it as a chance, and if it is more, lets use it in a responsible way.

General Reading

- Nanoscience and nanotechnologies: opportunities and uncertainties, *The Royal Society and The Royal Academy of Engineering*, July 2004.
- V. Balzani, M. Venturi and A. Credi, *Molecular Devices and Machines*, Wiley-VCH, Weinheim, 2003.
- S.A. Wolf, D.D. Awschalom, R.A. Burhman, J.M. Daughton, S. von Molnár, M.L. Roukes, A.Y. Chtchelkanova and D.M. Treger, Spintronics: a spin-based electronics vision for the future, *Science*, 2001, **294**, 1488.
- S.-Y. Wang and R. Stanley Williams, (eds), *Appl. Phys. A*, 2005, **80** (issue 6), 1133–1389.

References

1. Nanoscience and nanotechnologies: opportunities and uncertainties, *The Royal Society and The Royal Academy of Engineering*, July 2004.
2. R. Haag, *Angew. Chem.*, 2004, **116**, 280.
3. Q.A. Pankhurst, J. Conolly, S.K. Jones and J. Dobson, *J. Phys. D: Appl. Phys.*, 2003, **36**, R167.
4. K.E. Drexler, *Engines of Creation: The Coming Era of Nanotechnology*, Anchor Books, New York, 1986.
5. C. Phoenix and E. Drexler, *Nanotechnology*, 2004, **15**, 869.
6. M.C. Jimenez-Molero, C. Dietrich-Buchecker and J.-P. Sauvage, *Chem. Commun.*, 2003, 1613.
7. J.-P. Sauvage, *Chem. Commun.*, 2005, 1507.
8. A.M. Fennimore, T.D. Yuzvinsky, W.-Q. Han, M.S. Fuhrer, J. Cumings and A. Zettl, *Nature*, 2003, **424**, 408.
9. Y. Shirai, A.J. Osgood, Y. Zhao, K.F. Kelly and J.M. Tour, *Nano. Lett.*, 2005, **5**, 2330.
10. P. Miaudet, S. Badaire, M. Maugey, A. Derre, V. Pichot, P. Launois, P. Poulin and C. Zakri, *Nano. Lett.*, 2005, **5**, 212.
11. B.C. Regan, S. Aloni, R.O. Ritchie, U. Dahmen and A. Zettl, *Nature*, 2003, **428**, 924.
12. N. Hirokawa, *Science*, 1998, **279**, 299.
13. R.L. Cross, *Nature*, 2004, **427**, 407.
14. B. Bensaude-Vincent, *HYLE – Internat. J. Philos. Chem.*, 2004, **10**, 65.
15. A.K. Schmid, N.C. Barthlet and R.Q. Hwang, *Science*, 2000, **290**, 1561.
16. G. Moore, *Electronics*, 1965, **38**(8), 19 April.

17. R.P. Feynman, *Eng. Sci.*, 1960, **23**, 22; *J. Micromech. Syst.*, 1992, **1**, 60. www.zyvex.com/nanotech/feynman.html.
18. R. Bennewitz, J.N. Crain, A. Kirakosian, J.–L. Lin, J.L. McChesney, D.Y. Petrovykh and F.J. Himpsel, *Nanotechnology*, 2002, **13**, 490.
19. S.R. Forrest, *Nature*, 2004, **428**, 911.
20. B. Xu, X. Xiao and N.J. Tao, *J. Am. Chem. Soc.*, 2004, **126**, 5370.
21. Z.K. Tang, L. Zhang, N. Wang, X.X. Zhang, G.H. Wen, G.D. Li, J.N. Wang, C.T. Chan and P. Sheng, *Science*, 2001, **292**, 2462.
22. A. Bachtold, P. Hadley, T. Nakeshi and C. Dekker, *Science*, 2001, **294**, 1317.
23. J. Kong, N.R. Franklin, Ch. Zhou, M.G. Chapline, S. Peng, K. Cho and H. Dai, *Science*, 2000, **287**, 622.
24. P.G. Collins, K. Bradley, M. Ishigami and A. Zettl, *Science*, 2000, **287**, 1801.
25. Y. Wang, K. Kempa, B. Kimball, J.B. Carlson, G. Benham, W.Z. Li, T. Kempa, J. Rybczynski, A. Herczynski and Z.F. Ren, *Appl. Phys. Lett.*, 2004, **85**, 2607.
26. K.B.T. Teo, E. Minoux, L. Hudanski, F. Peauger, J.-P. Schnell, L. Gangloff, P. Legangeux, D. Dieumegard, G.A.J. Amaratunga and W.I. Milne, *Nature*, 2005, **437**, 968.
27. K. Terabe, T. Hasegawa, T. Nakayama and M. Aono, *Nature*, 2005, **433**, 47.
28. S.A. Wolf, D.D. Awschalom, R.A. Burhman, J.M. Daughton, S. von Molnár, M.L. Roukes, A.Y. Chtchelkanova and D.M. Treger, *Science*, 2001, **294**, 1488.
29. B. van Wees, *Nature*, 2005, **437**, 1249.
30. C.P. Weber, N. Gedik, J.E. Moore, J. Orenstein, J. Stephens and D.D. Awschalom, *Nature*, 2005, **437**, 1330.
31. Y.K. Kato, R.C. Myers, A.C. Gossard and D.D. Awschalom, *Science*, 2004, **306**, 1910.
32. J. Wunderlich, B. Kästner, J. Sinova and T. Jungwirth, *Phys. Rev. Lett.*, 2005, **94**, 047204.
33. M.C.K. Wiltshire, *Science*, 2001, **292**, 60.
34. R.A. Shelby, D.R. Smith and S. Shultz, *Science*, 2001, **292**, 77.
35. P.V. Parimi, W.T. Lu, P. Vodo and S. Sridar, *Nature*, 2003, **426**, 404.
36. H.-J. Eisler, V.C. Sundar, M.G. Bawendi, M. Walsh, H.I. Smith and V. Klimov, *Appl. Phys. Lett.*, 2002, **80**, 4614.
37. J. Wei, H. Zhu, D. Wu and B. Wei, *Appl. Phys. Lett.*, 2004, **84**, 4869.
38. *Nature*, 2003, **424**, 237 (Editorial).
39. G. Brumfil, *Nature*, 2003, **424**, 246.

Subject Index

Absorption cross section, 70
Adsorbate induced restructuring, 11
Adsorption hysteresis, 170
Adsorption isotherm, 168, 189
Aggregation, 153
Alloying, 252
Amphiphile, 153
Anharmonicity, 25
Anion-terminated surface, 145
Anisotropy energy, 104
Antiferromagnetism, 84

Backbending, 219
Band gap, 48, 51, 60
Band structure, 41
Bias voltage, 257
Bicontinuous structure, 155
Bilayer vesicle, 154
Bioaccumulation, 276
Biomimetics, 17
Blocking temperature, 90, 101, 103, 108
Bose-Einstein statistics, 120
Bragg equation, 16
Brillouin function, 82
Brunauer-Emmett-Teller equation, 163

Canonical ensemble, 218
Capillarity, 124
Capped nanocrystal, 214
Catalysis, 239
Catalysis, shape dependent, 246
Catalysis, size-controlled, 241
Catalysis under strain, 249
Chain, monoatomic, 106
Chain, one-dimensional, 107
Chemisorbed hydrogen atoms, 114
Clausius-Clapeyron equation, 128

Clausius-Mosotti equation, 54
Cluster, catalysis, 241
Cluster, core-shell, 156
Cluster, dynamics, 209
Cluster, electron affinity, 58
Cluster, icosahedral, 34
Cluster, ionisation potential, 56
Cluster, magnetic properties, 95
Cluster, melting, 178
Cluster, melting enthalpy, 209
Cluster, nucleation, 209
Cluster, phase transition, 209
Cluster, size, 1
Cluster, structure, 21
Cluster, superparamagnetism, 88
Coercive remanent magnetisation, 84
Coercivity, 104
Coexistence bands, 124
Cohesive energy, 34, 51
Colloidal gold, 70
Colloids, 25
Compensation effect, 247
Conduction electron ESR, 111
Configurational entropy, 150, 228
Confined nanofluids, 203
Confined water, 195
Confinement, electron, 63
Confinement, hole, 67
Confinement, strong, 62
Confinement, weak, 62
Contact angle, 125, 129, 133, 170
Contrast enhancement agents, 265
Cooperative interaction, 196
Cooperative phenomenon, 182
Coordination number, 6, 26, 34, 127, 192, 197
Core-shell structure, 65, 156

Coulomb blockage, 54
Critical micelle concentration, 154
Critical points, 124
Critical temperature, 175, 190
Cryoscopic freezing point depression, 223
Crystal growth, kinetic control, 139
Crystal morphology, 144
Crystallite formation, 136
Curie paramagnetism, 82
Curie-Weiss law, 93
Curie-Weiss paramagnetism, 83

Debye formula, 16
Debye relaxation, 197
Deep-trap emission, 67
Defect concentration, 150
Delivery of therapeutic drugs, 264
Density of states, 44, 47, 48
Dielectric catastrophe, 54
Dielectric properties, 194
Dielectric spectroscopy, 195
Diffusion under confinement, 198
Dimensionality, 42.96
Discrete periodic melting, 221
Discrete states, 41
Disorder, 265
Dispersion, 6
Divalent elements, 49
DLVO theory, 11, 25
Drug delivery, 264
Dubinin-Radushkevich equation, 168
Dynamic coexistence melting, 220
Dynamics in pores, 194
Dynamics of clusters, 209

Effective viscosity A153, 202
Einstein relation, 199
Electrical conductivity, 53
Electromigration, 268
Electron affinity, 51, 58
Electron confinement, 63
Electron paramagnetic resonance, 52
Electron spin resonance, 111
Electronic interaction, 252, 256
Electronic shells, 29
Electronic structure, 41, 60
Electron-phonon coupling, 73
Energy of adhesion, 133
Ensemble, canonical, 218

Ensemble, microcanonical, 219
Entropy of melting, 215
Entropy production, 121
Environmental hazards, 276
Ergodic theorem, 121
Ethical threats, 276
Evans-Polany relationship, 253
Excitons, 61
External bias voltage, 257

F-centers, 243
Fermi energy, 42, 96, 239, 249, 255
Fermi-Dirac statistics, 120
Ferrimagnetism, 84
Ferromagnetic domains, 106
Ferromagnetic exchange length, 106
Ferromagnetism, 84
Feynman, 2, 269
Field cooling, 106
Field-effect transistor, 45, 75, 271
Floater atom, 226
Fluctuation theorem, 120
Fluid phases, 124
Fluorescence, 65
Fluorescence quenching, 68
Frenkel-Halsey-Hill theory, 127
Friedel oscillations, 5
Frustration, 91

Geometric frustration, 93
Geometric shells, 26
Geometric structure, 21
Germs, 136
Giant magnetoresistance, 273
Gibbs phase rule, 123
Gibbs-Thomson equation, 129, 179, 221, 233
Glass transition, 236
Gradients near surfaces, 5

Hagen-Poiseuille law, 134
Health hazards, 276
Heat capacity, 215
Heat of adsorption, 165
Heisenberg model, 92
Hierarchically rough surface, 134
Hybridisation, 50
Hydrogen absorption, 111
Hydrogen induced melting, 222
Hydrophilic/hydrophobic switching, 135

Hydrophobic free energy, 154
Hydrophobic surface, 129
Hyperthermia, 264
Hysteresis, 104
Hysteresis loop, 170

Icosahedral clusters, 34
Immersion spectroscopy, 71
Interface, 5
Interfacial energy, 233
Interfacial tension, 178, 183
Inverse surface melting, 229
Ion conduction, 148
Ionisation potential, 50, 56
Irreversibility temperature, 85
Ising model, 92

Jahn-Teller distortions, 45

Kagome lattice, 94
Kawabata theory, 52
Kelvin equation, 125, 127, 170
Kinetic control, 263
Kinetic energy, 217, 224
Knight shift, 109
Knudsen diffusion, 200
Korringa relation, 53, 110
Kramers-Kronig relation, 71
Kubo gap, 47, 50, 96

Landau orbit, 46
Landauer constant, 271
Langevin function, 89, 104
Langmuir adsorption isotherm, 163
Laplace law, 13
Laplace-Young equation, 193
Latent heat, 210
Latent heat of fusion, 212
Lattice contraction, 12
Layering transition, 124, 185
Lennard-Jones fluid, 189
Lennard-Jones potential, 22, 166
Lindemann criterion, 209, 225
Liquid coexistence, 191
Liquid droplet, 128
Lotus effect, 129
Luminescence, shallow trap, 66

Magic numbers, 26, 29
Magnetic anisotropy, 87

Magnetic anisotropy energy, 90, 108
Magnetic exchange length, 101
Magnetic frustration, 92
Magnetic moment, gradient, 98
Magnetic properties, 81
Marangoni effect, 171
Mean square displacement, 200
Melting enthalpy, 209
Melting mechanisms, 219
Melting of thin layers, 233
Melting point, 209
Melting point depression, 129, 182
Melting temperature, 216
Meniscus curvature, 163
Mesopore volume, 169
Metal cluster, 60
Metal-support interaction, 255
Metal-to-insulator transition, 55
Meyer-Neldel rule, 247
Microcanonical dynamic temperature,
    224
Microcanonical dynamical equipartition
    postulate, 223
Microcanonical ensemble, 219
Microfibre fabrics, 134
Microstrain, 11
Mie theory, 71
Military application, 277
Miscibility gap, 192
Molecular magnet, 86
Molecular manufacture, 266
Monodispersity, 139
Moore's law, 269
Morphology, 148
Morse potential, 22
Mott transition, 47

Nakaya diagram, 142
Nanodevice, 266
Nanoelectrical device, 74
Nanoelectronics, 269
Nanomachine, 268
Nano-magnetism, 272
Nanomaterial, 1
Nanomechanics, 267
Nanomotor, 269
Nanoparticle, 1
Nanoporous membranes, 265
Nanoscale mass conveyor, 267
Nanoscopic ensemble, 119

Nanosolution, 193
Nanoswitch, 271
Nanotube filament, 275
Nanowire, 271
Negative heat capacity, 217
Negative refractive index, 274
Newtonian fluid, 134
Newtonian viscosity, 202
NMR-cryoporometry, 184
Non-collinear spin structure, 93
Nonmetal-to-metal transition, 47
Nuclear Magnetic Resonance, 53, 109
Nucleation, 209
Nucleation rate, 138
Nucleation theory, 136

Odour eaters, 242
Optical properties, 69
Organic light-emitting devices, 270
Ostwald ripening, 16, 140
Oxidation overpotential, 246

Pair correlation function, 188
Particle size, 11
Passivating layer, 263
Pauli Paramagnetism, 42, 90
Periodic table, 33
Phase coexistence, 123, 229
Phase concept, 122
Phase diagram, 187
Phase rule, 123
Phase separation, 192
Phase transition, 123, 209
Phase transition enthalpy, 129
Phase transition temperature, 175
Photocatalyst, 266
Photochemical process, 65
Photoelectron spectra, 52
Photoluminescence, 66
Photonic crystal, 273
Photophysical process, 65
Polarisation energy, 50
Pore condensation, 163
Pore critical temperature, 176
Pore criticality, 170
Pore size, 169
Potential, energy, 217
Potential, confining, 32, 62
Potential, Lennard-Jones, 22, 166
Potential, Morse, 25

Pressure-induced phase transition, 193
Product-blocked surface, 253
Pseudo-atom, 29, 43
Pseudomorphic growth, 235
Public awareness, 277
Public dialogue, 277

Quantum bits, 272
Quantum computing, 272
Quantum conductance, 272
Quantum dot, 42
Quantum dot electronic device, 74
Quantum effect, 2
Quantum layer, 42
Quantum mirage, 47
Quantum size effect, 61, 245
Quantum statistics, 120
Quantum tunnelling of magnetisation, 88
Quantum wire, 42
Quasi liquid layer, 142
Quasi-crystalline, 21, 48

Radial gradient of magnetic moments, 98
Root mean square fluctuation, 225

Saturation magnetisation, 101, 107
Scalable effects, 2
Scaling theory, 192
Scanning tunnelling microscope, 3
Scherrer formula, 11
Schottky contact, 255
Second law of thermodynamics, 120
Self-assembly, 18, 152, 270
Self-cleaning property, 134
Self-coiling, 148
Self-diffusion coefficient, 199
Self-replicating robots, 267
Semiconductor cluster, 60
Sensing technology, 277
Sensor technology, 266
Shape control, 141
Shape dependent catalysis, 246
Shape-complementarity, 152
Shear stress, 202
Shell, electronic, 29
Shell, geometric, 26
Shell periodicity, 27
Single file diffusion, 200

Single-molecule conductance, 270
Single-molecule nanocar, 267
Single-molecule relaxation, 196
Sintering, 141
Size-controlled catalysis, 241
Size-distribution focusing, 141
Size-scaling effect, 222
Skin depth, 111
Social threats, 276
Solid core - liquid shell model, 210
Spatial constraint, 167
Specific heat, 44
Spherical micelle, 154
Spill-over, 71
Spin current, 273
Spin glasses, 91
Spin Hall effect, 273
Spin liquids, 91
Spintronics, 272
Spontaneous magnetisation, 85
Stern-Volmer quenching, 68
Stirling's approximation, 120
Strain free energy, 233
Strong confinement, 62
Structural phase transitions, 233
Structure directing agent, 196
Super-atom, 2, 31, 59
Super-atom clusters, 31
Supercooling, 138
Superhydrophobicity, 130
Superlattice, 156
Superparamagnetism, 88
Supersaturation, 137
Supramolecular chemistry, 18
Surface adsorption, 163
Surface alloy phase, 236
Surface curvature, 125
Surface fluxionality, 265
Surface free energy, 132
Surface, intelligent, 265
Surface premelting, 228
Surface reconstruction, 9
Surface restructuring, 239
Surface roughening, 252
Surface roughness, 130
Surface segregation, 10
Surface state, 64
Surface tension, 14, 125

Surface-to-volume ratio, 6, 27
Symmetry, 42, 48, 96

Temperature, local, 122
Template synthesis, 144
Thermocapillarity, 171
Thermodynamic factor, 199
Thermodynamic stability, 263
Thermoelectric potential, 255
Thin film, 106
Thin film solar cell, 274
Time averaged kinetic energy, 224
Toxicity, 276
Transport property, 201
Triple point capillary condensation, 190
Tuning electronic properties, 242
Turnover frequency, 253

Under-potential deposition, 234
Usefulness, 275

van der Waals cluster, 29
Van Vleck paramagnetism, 91
Vapour pressure, 128
Velocity autocorrelation function, 229
Viscosity under confinement, 198
Vogel-Fulcher-Tammann relationship, 195
Volcano-shaped plot, 253
Volume free energy, 233

Weak confinement, 62
Wealth creation, 275
Weimarn's law, 139
Wenzel's law, 130
Wilson transition, 42
Work function, 50, 55
Wulff's law, 140
Wulff's theorem, 14

X-ray magnetic circular dichroism, 109
X-Y-model, 92

Young equation, 124, 181
Young-Laplace equation, 125, 178

Zero-field cooling, 106